How Not to Network a Nation

Information Policy
edited by Sandra Braman

The Information Policy Series publishes research on and analysis of significant problems in the field of information policy, including state-law-society interactions as well as decisions and practices that enable or constrain information, communication, and culture irrespective of the legal siloes in which they have traditionally been located. Defining *information policy* as all laws, regulations, and decision-making principles that affect any form of information creation, processing, flows, and use, the series looks at the formal decisions, decision-making processes, and entities of government; the formal and informal decisions, decision-making processes, and entities of private- and public-sector agents that are capable of affecting the nature of society; and the cultural habits and predispositions that support and sustain government and governance. The parametric functions of information policy at the boundaries of social, informational, and technological systems are of global importance because they provide the context for all communications, interactions, and social processes.

How Not to Network a Nation

The Uneasy History of the Soviet Internet

Benjamin Peters

The MIT Press
Cambridge, Massachusetts
London, England

First MIT Press paperback edition, 2017

© 2016 Massachusetts Institute of Technology

This book was set in Stone by the MIT Press.

Library of Congress Cataloging-in-Publication Data
Names: Peters, Benjamin, 1980- author.
Title: How not to network a nation : the uneasy history of the Soviet internet / Benjamin Peters.
Description: Cambridge, MA : MIT Press, [2015] | Series: The information policy | Includes bibliographical references and index.
Identifiers: LCCN 2015038371 | ISBN 9780262034180 (hardcover : alk. paper) ISBN 9780262534666 (pb.)
Subjects: LCSH: Computer networks—Soviet Union—History. | Internetworking (Telecommunication)—Research—Soviet Union—History.
Classification: LCC TK5102.3.S68 P48 2015 | DDC 384.30947/09045—dc23
LC record available at http://lccn.loc.gov/2015038371

Joli Jensen, Michael Schudson, Fred Turner, Gary Browning
Four mentors at four schools

Contents

Contents

Series Editor's Introduction

Sandra Braman

In this first-ever book-length treatment of early Soviet intelligent network design history, Benjamin Peters uses uncanniness as a method. He makes use, that is, of the disorientation that results when the familiar is encountered in an unfamiliar context, broadening and deepening what we believe that we know about the familiar. This can be a dangerous endeavor. Avatar designers and others fear the "uncanny valley"—where the nonhuman is so close to the human that the difference cannot be discerned—because that is literally too close for viewer/user comfort. "That's different," they say in those cultures with traditional concerns about trolls, those who look like people but in fact are not.

Historically, the uncanny "other" was supernatural and not necessarily to be trusted with matters of this world. For Peters, the "other" is the Soviet Union. What he found is based on original multilingual archival research and oral interviews with those who were involved in the design processes. Peters describes what he sees as the capitalist features of the Soviet world that undermined its networking efforts and what he views as the socialist characteristics of the United States that produced the Internet. On the face of it, this suggests deep contradictions within capitalist and socialist systems that belie the claimed and apparent differences between the two blocs. But even those concerned with cybersecurity acknowledge that it can be difficult to identify the "other" in the network environment. What was the uncanny valley in this analytical zone?

For those who think about the information economy, differences between the East and West are indiscernible. Cristiano Antonelli's (1992) seminal insights into the nature of the information economy, in which cooperation and coordination are as important as—or more important than—competition for long-term economic success were inductively developed from detailed studies of the practices and activities of transnational corporations on both sides of the iron curtain (many funded by the unfortunately short-lived United Nations Center on Transnational Corporations). What Peters presents as counterintuitive actually provides further evidence of the transition to a global information economy in which ideological differences may still provide motivations but not explanations. Work of this kind, which looks across political environments, is particularly valuable as we struggle to make policy for a world in which network politics is genuinely global even though state-centric geopolitical distinctions remain.

Theoretical pluralism has been familiar since the 1980s, but on reading Peters one suddenly realizes that most of those who take such an approach tend to prefer particular types of causal probability even as they roam across theories and disciplines. Peters is not only interdisciplinary but also travels across the levels and qualities of the likelihood that any given causal factor will be determinative in a given circumstance. In this history of early Soviet network design efforts, Peters ranges from unpacking institutional rigidities that did successfully shape knowledge production and use to focusing attention on contingencies that can radically affect ultimate outcomes. His heterarchical approach to policy analysis importantly reminds us of the need to examine the interplay among decision-making processes as well as among players. And Peters returns again and again to the centrality of ideas in policymaking, devoting a full chapter to the history of cybernetics in the Soviet Union during the period covered.

Oddly, according to the *OED*, the notion of the "uncanny", came into written use a century before the word "canny" was seen. This may be an artifact of the processes by which materials survive, but it is still interesting. Peters's multilingual archival research and oral interviews with individuals involved in the Soviet efforts have yielded a picture of network conceptualization and decision-making processes fascinating not only in their own right but also for what they offer to those who study and live with intelligent networks in other parts of the world. We *are* the other, in the global network. With this book, Peters deepens our ken of networks—a fundament of information policy since at least the 1830s and the telegraph—and brings their study into the next generation.

Prologue

The seeds of this book were first planted as I stood on the left bank of the Volga River in Balakovo, Russia, one evening in the spring of 2001. Balakovo, where I was living for several months doing volunteer service, was a pleasant city of roughly 200,000 people who were struggling through the economic depression that was sweeping Russia's rust belt. As I reflected on my picturesque surroundings—green trees, rolling hills, and the setting sun's reflections on the river—I sensed that something was out of place. The peculiar features that were visible on the horizon, backlit by the sunset, belonged to the Saratov hydroelectric dam, one of the world's hundred largest dams by output and stretching over 1,200 meters in length to form the enormous Saratov reservoir. The city of Balakovo also is home to a thermal heat power plant and a nuclear power plant with four working nuclear reactors (the construction of two other nuclear plants was suspended in 1992). If local rumors were to be believed, Balakovo once boasted secret Soviet military factories, one of which produced a material for the cosmonautic industries that was so tough that napalm balled up and rolled off it. This peculiar pairing of natural scenery and outsized industrial infrastructure struck me on the riverbank that evening. What force of imagination and statecraft, I puzzled, could have decided to graft such hulking industry onto such a remote city—and why would it do so? Thus began my interest in the outsized infrastructural imagination of Soviet planners.

Six years later, in 2007, those seeds sprouted into the question driving this book. As a doctoral student at Columbia University, I wanted to learn more about the international sources of the information age, a topic that first crystallized for me in Fred Turner's graduate seminar on "Computers, Information Ideology and American Culture since World War II" at Stanford University in 2005. If, to gloss Whitehead, all philosophy begins as a series of footnotes to Plato, then this book began with an obscure footnote in Flo Conway and Jim Siegelman's popular biography of Norbert Wiener. As I was rereading the book's references one evening in 2007, I stumbled on a passing reference to a declassified, Freedom of Information Act–recovered 1962 Central Intelligence Agency report about a new Soviet initiative to develop a native "unified information network."[1]

That footnote triggered a question that was so tenacious that I had to write this book to shake it: why were there no Soviet developments comparable to the ARPANET in the 1960s? It made sense that, at the height of the cold war technology race, Soviet cyberneticists would try to build a "unified information network"—and yet I knew nothing about their efforts or outcomes. I was hooked. What had happened? Why was there no Soviet Internet?

Over the next eight years, the question drew me to archives and interviews in Moscow and Kiev. After spending a year exhausting the available leads, literature, and FOIA requests available from New York, I traveled to begin archival work in Moscow, although initially this proved a dead end. Marshall McLuhan once quipped that the first thing a visitor needs to know about Russia is that there are no phonebooks.[2] His point is that a foreigner in Russia needs to have contacts already in place. (Or as the Finns say: in Finland, everything works and nothing can be arranged. In Russia, nothing works but everything can be arranged.) And so, with all the tools but none of the social network, I found myself shuffling through dusty documents that were lit by a single flickering light bulb in Moscow archives. Then in 2008, good fortune smiled when, while chasing down references to Nikolai Fedorenko and Viktor Glushkov in Moscow, I began a correspondence with the Massachusetts Institute of Technology historian Slava Gerovitch, who emailed me from Cambridge a draft of his article "InterNyet: Why the Soviet Union Did Not Build a Nationwide Computer Network" that became the basis for this book.[3] Gerovitch also put me in touch with key contacts in Kiev, and my rapidly expanding social network led to dozens of interviews and contacts, out-of-the-way archives (including stacks of papers in the closet of an abandoned office), and unprecedented access to historical materials over years of research and writing. On the surface, this book is about why certain computer networks did not work in the Soviet Union, but the story turns on the basic fact that social networks in the region have long operated according to their own rhythms and reasons.

Writing this book has proven to be a valuable learning process. When I set out in 2007 to study early Soviet networks, I had a vague sense that the resulting scholarly work would intersect media and communication studies, the history of science and technology, and social thought that informed information policy discussions, but I did not anticipate the work in institutional and historical economics, the sociology of economics, and organizational theory that the story required. Least of all did I imagine that this story would throw me headlong into a study of Soviet bureaucracies. It is my hope that this work will lighten some of that burden for the patient reader. In the end, this book should be understood primarily as an interdisciplinary work of synthesis driven by a fascination with the relationship between communication technology and people. I have tried to write for the media and technology scholar as well as the general-interest reader, although the book draws on history, area studies, and social commentary to inform the emergent subfield of network studies in information policy as well. Like all these fields, its primary orientation is not to a single discipline but to the scholarly enterprise of making strange modern network culture, a technique that the Soviet critic Viktor Shklovsky first popularized as *ostranenie*, or "defamiliarization."[4] It seeks to offer what historian Peter Brown calls "salutary vertigo" or a disorientation that clarifies the foreignness of a modern networked culture that was once thought familiar.[5] To do so, this work seeks to separate readers from hidden assumptions about modern networks and the social, technological, political, and economic conditions that organize and are subsequently organized by it. For me, this book began as an essay on the forgotten origins of computer networks in the Soviet Union and ended up being about much more, including a cautionary tale in the annals of technological innovation and a critical reflection on the assumptions steeping the current network world.

Introduction

There is much which we must leave, whether we like it or not, to the un-"scientific" narrative method of the professional historian.

—Norbert Wiener, *Cybernetics, or Control and Communication in the Animal and the Machine*, 1948, concluding line

The Soviet Union was home to hidden networks. The story told here about those networks hangs on a hook that is unfamiliar to most readers and scholars—the Soviet Internet. At first glance, pairing the Internet and the Soviet Union appears paradoxical. The Internet first developed in America and became popular only after the Soviet Union collapsed. The Internet suggests to general readers open networks, flat structures, and collaborative cultures, and the Soviet Union signals censored networks, hierarchies, and command and control cultures. What, then, could the phrase *Soviet Internet* possibly mean?

The central premise of this book holds that there was once something that we might think of as the Soviet Internet. Between the late 1950s and the late 1980s, a small group of leading Soviet scientists and administrators tried to develop a nationwide computer network that was designed for citizen communication and sweeping social benefits. This book is about their story. At the height of the cold war technology race, the Soviet Union was awash in intelligence about contemporary Western initiatives, including the Semi-Automatic Ground Environment (SAGE) project at the U.S. Department of Defense. The Soviet state had all the necessary motives, mathematics, and means to develop nationwide computer networks for the benefit of its people and society. This book also ventures analysis on why, despite pioneering national network projects from the most promising of scientists and administrators, the Soviet state proved unable and unwilling to network its nation.

This much is clear: the Soviet Union never had the Internet as it is known today.[1] Rather, in the early 1960s, Soviet cyberneticists designed the most prominent of the network projects examined here—the All-State Automated System (OGAS)—with the mission of saving the entire command economy by a computer network. Their elaborate technocratic ambition was to network, store, transmit, optimize, and manage the information flows that constituted the command economy, under the guidance of the Politburo and in collaboration with everyday enterprise workers, managers, and planners nationwide.

The historic failure of that network was neither natural nor inevitable. Its story is one of the lifework and struggles of often genius cybernetic scientists and administrators and the institutional settings that were tasked with this enormous project. The question deserves a sympathetic and rigorous examination of the Soviet side of the story. Why did Soviet networks like the OGAS not take root? What obstacles did network entrepreneurs face? Given unprecedented Soviet investments and successes in mathematics, science, and some technology (such as nuclear power and rocketry), why did the Soviet Union not successfully develop computer networks that were capable of benefiting a range of civilian, economic, political, social, and other human wants and needs? How might we begin to rethink our current network world in light of the Soviet experience?

I propose that the primary reason that the Soviets struggled to network their nation rests on the institutional conditions supporting the scientific knowledge base and the command economy. Those conditions, once examined, challenge conventional assumptions about the institutions that build open, flat, and collaborative networks and thereby help recolor the cold war origins of the information society. It is a mistake, as the standard interpretation among technologists and some scholars have it, to project cold war biases onto this history. Our networked present is the result of neither free-market triumphs nor socialist state failures.

That said, let us begin with a slight twist on the conventional cold war showdown: the central proposition that this book develops and then complicates is that although the American ARPANET initially took shape thanks to well-managed state subsidies and collaborative research environments, the comparable Soviet network projects stumbled due to widespread unregulated competition among self-interested institutions, bureaucrats, and other key actors. The first global civilian computer networks developed among cooperative capitalists, not among competitive socialists. The capitalists behaved like socialists while the socialists behaved like capitalists.

In the process of examining and elaborating on that plain statement about the cold war history of networks, this book describes two intersecting approaches to larger questions of social control and change—one institutional and the other technological. The first approach looks at the context of Soviet institutions and political bodies that were preoccupied with both the paperwork and the power brokerage behind the socialist command economy. The question of how to organize economies, especially but not only the Soviet command economy, is shown to be political before it is economic. The second approach accounts for the attempts of Soviet cyberneticists to build a computer system over a period of about thirty years from 1959 to 1989 that would control in real time the economy's problems. The two approaches— political economy and computing technology—combine and play out here on the common stage of Soviet cybernetics, a midcentury discipline that was interested in systematizing all organization problems with computing technology. The result is a tragic story that addresses questions that are central to the history of technology and global media theory: what makes the same technology take shape differently in different contexts?

To explore that tragedy, the book sets up the dramatic potential of a networked command economy, the loss of that potential in the hands of the state, and a critical reclamation for contemplation, reflection, and contemporary instruction. The limitations of this work's scope are also clear. Although it focuses primarily on the cybernetics and economic concerns besetting Viktor Glushkov and his Kiev-based OGAS team between 1959 and 1989, the setting is broader, including the military, industrial, and academic complexes that stretched from the seat of power Moscow to other cities, including St. Petersburg to the north and Akademgorodok (a science city that was nestled deep in Siberia) three thousand kilometers to the east.[2] The book also seeks to comment on the Soviet Union as a perceived state of exception on the global geopolitical stage. As one pole in the global cold war, the Soviet state stood unrivaled among socialist states in terms of international military and political influence.[3] In their search for a balance of focus and breadth, historians of science and technology have called for midpicture history, or a case study drilled deeply to explore intersecting historical subdisciplines (not entirely unlike Robert Merton's middle-range theory). This book is not a midpicture history, although I hope its best moments may model how media history and theory can move in tandem with information science and technology. In its most ambitious moments, this book offers a synthesizing commentary (in the premodern sense of the central genre of scholarship, not derivative status) about the sources of the modern network age.[4]

This book seeks to complicate the popular memory of the Soviet Union— its heady promises of socioeconomic justice as well as its parade of horribles, including authoritarian abuses, violence, and a cumbersome state hierarchy that subjected its citizens to political oppression and information censorship. It examines the Soviet command economy, which proved inflexible to the fluctuating demands of the emerging global network economy and eventually imploded on itself. Some readers may feel that the Internet and the Soviet Union seem to be fundamentally opposed information projects: one is a salvific vehicle for the invisible hand of modern-day commerce, and the other is remembered for its dead hand; one led to the knowledge explosion that is *Wikipedia*, and the other, to the nuclear catastrophe at Chernobyl; one produced Linux, and the other, the Lada; one is a haven for technoenthusiasts, libertarians, and free-speech absolutists, and the other, the whipping boy for the same. But I seek to bring to English-language readers the story of the Soviet computer network in its own terms. Given that the story is not singular, my emphasis is on relating the untold story of the All-State Automated System project and its research network led by the mathematician Viktor Glushkov in Kiev (the current capital of Ukraine) between 1959 and 1989. The case study arrives couched in commentary that seeks to upend and move beyond residual binary narratives about the cold war origins of the current networked age.[5]

The internal historical setting for the tragic tale begins with the turbulent grab for power that followed the death of Joseph Stalin in 1953 and stretches through the halting internal unraveling of that power in the 1980s. There was an unusual contender for filling the political vacuum left by Stalin's passing. To the scientists under study here, Stalin's best replacement was no person at all but rather a technocratic conviction that computer-aided governance could avoid the past abuses of its strongman state. The All-State Automated System was a utopian vision of a distinctly state socialist information society as well as, closer to home, a familiar story of how bright men and women struggle to employ both might and machines in the service of social justice and greater public goods.[6]

A thin line sometimes separates tragedy from comedy. Backlit by reflections on cold war political economic orders, the fickle muses of historical contingency staff this drama. For example, family preferences for warmer weather ended up shifting the centers of scientific development, empty chairs at crucial meetings sank decade-long campaigns, informal whims of power shipwrecked careers and perhaps countries, basic notational systems (not sophisticated algorithms) revolutionized long-term strategic thinking (and Soviet chess), and countless other details rained down via informal bureaucratic actions on the Soviet knowledge base. All these and others blur the comic and tragic elements. The Soviets could have developed a network contemporary to the ARPANET, and yet they did not. What makes this story tragic is not that the Soviet political, economic, and technological networks collapsed but that the deeper problems that beset the USSR have been transformed but have not disappeared. The twenty-five years following the collapse of the Soviet Union have reaffirmed that Russia, although no longer in a superpower showdown with the West, remains anything but a negligible actor on the global stage and that the patterns of its state governance are much older than the post-Soviet transition. By triangulating across the central Soviet-American cold war axis to emphasize Ukrainian and other liminal people and places, this book aims to help readers rethink residual cold war misunderstandings in popular network and digital media discourse while simultaneously showcasing the institutional tensions at the heart of modern-day networked practices, policies, and polities.

The curtain parts on two anecdotes about Soviet networks. The first introduces the central story, and the second marks the limit of that story. In late September 1970, a year after the ARPANET went online, the Soviet cyberneticist Viktor Glushkov boarded a train from Kiev to Moscow to attend what proved to be a fateful meeting for the future of what we might call the *Soviet Internet*. On the windy morning of October 1, 1970, he met with members of the Politburo, the governing body of the Soviet state, around the long rectangular table on a red carpet in Stalin's former office in the Kremlin. The Politburo convened that day to hear Glushkov's proposal and decide whether to build a massive nationwide computer network for citizen use—or what Glushkov called the All-State Automated System (OGAS, obshche-gosudarstvennyi avtomatizirovannaya system), the most ambitious computer network project of its kind in the world at the time. OGAS was to connect tens of thousands of computer centers and to manage and optimize in real time the communications between hundreds of thousands of workers, factory managers, and regional and national administrators. The purpose of the OGAS Project was simple to state and grandiose to imagine: Glushkov sought to network and automatically manage the nation's struggling command economy.

What transpired in Stalin's former office that day enters into the story. Throughout this (and perhaps all) history, the messy details often matter most. In this case, two crucial chairs in that committee room were empty on that particular day due to the contingencies of the calendar and competitive bids for power. This book's analysis will note how pesky details often reveal hidden patterns of institutional (mis)behavior that structure and reshape the interests of public actors, organizations, and even economic and social relations. Taken together, the history and analysis of the OGAS and related attempts to network and command the Soviet economy tell a story with consequences for the history of cold war computer networks and our understanding of the current networked world that emerged from the cold war itself.

The second anecdote took place not far from Glushkov's fateful meeting in the Kremlin. Here, in a top-secret chamber in a cement bunker, or *shariki* ("spheres" or "globes"), buried deep underground somewhere outside of Moscow, was a very different kind of computer network. In that small room, a few uniformed personnel sat before flickering computer screens that were powered by an independent generator purring audibly nearby but out of sight. The single closed door was of reinforced metal with a self-locking mechanism, and behind it a long ladder ascended into a network of underground tunnels overhead. The chairs were bolted to the floor and pivoted to allow the military officers to review a control panel lined with information displays—satellite data and security camera feeds, telephone and radio signals, Geiger counters and seismographs, and other instruments for measuring the world above. These men sat at their consoles, operating as cogs in a larger sociotechnical machine. They were trained so that if or when the time arrived, they would observe the sensors, orient and input certain coordinates and a timetable, flip switches, and press a button that would end the world in a nuclear Armageddon.

This is Dead Hand, the semiautomatic nuclear-defense perimeter system that was first installed in the late Soviet Union. The details above are mostly pure invention, and yet the network system is real. Formally called Systema "Perimetr," the perimeter system was imagined under Brezhnev as a fail-deadly deterrence mechanism for ensuring second-strike capacity in the nuclear cold war.[7] These men—not unlike the U.S. workers who staffed the Emergency Rocket Communication System from 1961 to 1991—sat in the top-secret underground command-and-control center of their nation's perimeter system. The data were fed into computer consuls to confirm whether the enemy had struck first. If an American military strike effectively disabled the regular Soviet command-and-control military leadership above ground from swiftly retaliating, then the strategy maintained that the Dead Hand would stand ready to trigger "a spasm of destruction."[8] After the national computer network system was activated, it would put on alert nuclear-tipped intercontinental missiles that were stored thousands of miles away. The red button, once pressed, would launch a massive retaliatory nuclear strike, enacting swift revenge at a global cataclysmic scale. Behold the apocalypse—delivered by national network.

This book is about civilian networks, not military networks. This is a deliberate choice. I choose to emphasize public networks because a network built for every Soviet worker still speaks to the popular and scholarly imagination of our current socially networked world in ways that closed military networks do not, although, as we will see, the military's relationship to technological innovation backlights the whole stage of cold war science.

A sideways look at some of the discourse about online commerce today proposes the enduring relevance of the Soviet socialist revolution that was consummated a century ago. Both the Internet and the Soviet command economy promise the revolutionary realization of the means for socialist or collectivist production on a mass scale. In the rhetoric of networking collective consciousness and crowd-sourced collaboration, we see the unlikely alliance of *Wired* editor Kevin Kelly's hive mind, open-source software promoter Eric Raymond's bazaar, and Marxist revolutionary Leon Trotsky's collective farm.[9] Long before Internet enthusiasts were around, Soviet enthusiasts were promising that workers (users) could meet the needs of the masses (crowds) through collective modes of resource sharing and collaboration (peer-to-peer production).

Few, if any, contemporary scholars recognize these concerns as fundamental to our modern network culture, and yet they persist in coloring views of both past and future. This is no accident: the concurrent emergence of cyberspace and post-Soviet affairs entered scholarly and popular discourse at the tail end of the previous century. For example, sociologist Manuel Castells has developed an extensive argument detailing how the Soviet Union failed to enter the information age, which this book is in some ways a sideways response to, and legal scholar Lawrence Lessig used his experience observing the rapid deregulation and privatization in post-Soviet economic transition in the early 1990s as a formative analog for what he felt was an equally disastrous attitude about the supposed unregulability of cyberspace common in the late 1990s.[10] Since then, scholars have recognized that the summary experiences of perhaps the last two great information frontiers of the twentieth-century—the rise of post-Soviet economic transition and the Internet—present not, as Francis Fukuyama infamously claimed, the end of history so much as a new chapter in it. Leading cyber legal scholar Yochai Benkler has argued for a middle way by observing how online modes of "commons-based peer production" sustain capitalist profit margins through collectivist forms of reputational altruistic communities that do not depend on individual self-interest.[11] From the final chapters of Soviet history, we may begin to observe and puzzle through the perennial fact that, for many Western technologists and scholars, the promise of socialist collaboration shines brightest online today—a promise that the Soviet OGAS designers were among the first to foresee.

None of the conditions—technological, sociological, economic, or otherwise—for the flourishing of computer networks are necessarily as we may think. As Melvin Kranzberg's first law of technology holds, technology is neither positive, negative, nor neutral.[12] The same holds for society and economy. By looking at failed network projects, I seek to flip science anthropologist and philosopher Bruno Latour's aphorism that technology is society made durable. We observe in the collapse of the Soviet network projects a lesson for humans who live in a fragile world: society too is technology made temporary.[13] The Soviet experience with networks reminds us that although computer networks are prospering today, our modern social assumptions about those networks are no more inevitable or permanent than those of the Soviets. Our current beliefs about networks will pass. This book looks to take in a new direction what science and technology scholars Geoffrey C. Bowker and Leigh Starr have called an "infrastructural inversion": looking closely at the alternative setting of a Soviet networked society can shake up a modern mental infrastructure that makes the current networked environment appear natural and necessary.[14] Sometimes the best way to see something is to look away from it. The French revolution, as historian Eric Hobsbawm has noted, did not become *the* French revolution until it was seen in the context of the British industrial revolution and the revolutions of 1848.[15] We stand to apprehend the current network transformations better by placing the past in the context of a wider world. By exploring the pathway that was once taken and then abandoned in cold war networks, I hope to help unsettle, broaden, and deepen our imagination for the possibilities that gave rise to the modern networked media environment.

The literature on which this book builds is growing. Above all, this book builds on the historical foundation that was laid by the pioneering works of historians of Soviet science, Slava Gerovitch and Loren Graham.[16] Slava Gerovitch's article "InterNyet: Why the Soviet Union Did Not Build a Nationwide Computer Network," which he shared with me in draft form while I was independently pursuing the Soviet Internet story in archives in Moscow, jumpstarted this history with a treasure chest of scholarly leads. His work has opened many windows into the Soviet history of science and its associated social problems. The literature in English on the midcentury development of computer networks—by leading scholars such as Janet Abbate, Finn Burton, Paul N. Edwards, Fred Turner, and Thomas Streeter—also includes works that examine the creative communities, institutional innovation and setbacks, cold war tensions, and Western internal politics that backlight this particular case study.[17] This work attempts to help internationalize the core insights of this sociologically sensitive body of analysis into the people and places that shape networks.

The literature also teaches that the significance of the global spread of a social network often precedes, exceeds, and coevolves with that of any specific technological network. To borrow a line from Elihu Katz, international communication networks precede national computer networks.[18] Along these lines, historian of technology Eden Medina's *Cybernetic Revolutionaries: Technology and Politics in Allende's Chile* advances a seminal history and analysis of early technological and political attempts to network another socialist state during the cold war. Her close and careful analysis of the people involved in the creation of Project Cybersyn (especially 1971 to 1973) reveals how the significance of technological projects carries beyond and exceeds that of specific network projects.[19] Her work, together with other recent scholarship on international cybernetic movements, helps outline the central cast of characters in this book.[20] This cast was not selected exclusively from cybernetic scientists or administrators. Rather, the characters are drawn from what I call the "knowledge base" of the Soviet Union—theoretical and applied scientists, their laboratories and research centers, students in universities, administrators in the academies of science, state office bureaucrats, generals in the Ministry of Defense, ideologues and censors in the scholarly and public press, the secret police, functionaries, officials, midlevel managers, members of the Central Committee of the Communist Party, and others whose careers depended on the management, manipulation, and representation of knowledge as an intellectual, institutional, and innovative product.[21]

Finally, a practical note about language. All translations from Russian and Ukrainian into English are my own unless otherwise noted. In translating, as Stephen Jay Gould says, "we reveal ourselves in the metaphors we choose for depicting the cosmos in miniature."[22] This is true of the translation process as a way of trying to bring separate languages into resonance. Sometimes words can be translated straightforwardly. For example, this work, an interdisciplinary exercise in the emerging field of network studies, seeks to articulate a fluid discourse around the central term *network*. The term *network*, like other keywords in digital discourse, packs more meaning than is usually seen and has roots in the textile industry of lacework, like the Jacquard loom behind computer programming techniques (there may be more silk than silicon to the information age). The Russian term *set'* maps fairly well onto my three English uses of the term *network*—(1) a

technical communication network understood as interlinked digital, electronic, telephonic, or other channels of communication; (2) the complex
sociotechnical assemblage of heterogeneous relations that link people,
institutions, and the administration of markets, states, and other actors in
everyday life; and (3) an abstract organizational mode that maps the linkages between any set of objects, such as graph theory in mathematics.[23]
Although all of these meanings are in play here, what we assume to be a
relatively settled term today behind the concept of *network* (*set'*) took up in
Soviet discussions an even wider set of terms such as *base, complex, cluster,*
and most characteristically for computers connected over distances, *system.*

At other times, Russian terms reveal their own world in how they resist
easy translation. I occasionally retain, for example, the early Soviet term for
computer, "the automatic high-speed electronic calculating machine" (*avto-
maticheskaya byistrodeistvuyushchaya elektronicheskaya schyotnaya mashina*
and its various shortenings) for its splendidly descriptive bulk that signals
perhaps the most elegant definition of new media I know: new media are
those media we do not yet know how to talk about.[24] The probability theorist Aleksandr Ya. Khinchin revealingly renders what is known in English as
"queuing theory" (used by information theorists to describe how data packets wait in line) as "mass-service theory" (*teoria massovogo obsluzhivaniya*) in
Russian.[25] Sustaining the anthropological gaze requires depicting the variable sets of cultural, social, and political values in comparative relief with
the network elements that are all too familiar in modern culture, which I
have attempted to do here whenever relevant.

I have also tried to write with the conviction that plain language packs in
its own insights. By proposing for further examination that the first global
civilian networks took shape thanks to capitalists behaving like socialists,
not socialists behaving like capitalists, I understand the terms *capitalism*
and *socialism* in the ordinary way. I define *capitalism* as the order of the
market economy, where economic actors act independent of the state, private property rights are reasonably secure and dominate most enterprises,
prices and trade are predominantly free, state subsidies are limited, and
transactions mostly monetized. *Socialism,* by contrast, is an economic order
of the command economy where the opposite can usually be expected,
although with its instinct to communism operating according to the moral
and political principle "from each according to their abilities, to each
according to their needs."[26] The argument here depends not on collapsing
that definitional divide but on revealing how that ordinary understanding
falls short of describing mixed constellations of competitive and collaborative practices—public-private and state-market formations that belie and

tweak our sense of these opposing economic orders. Evidence complicates the tidiness of ideas. This is a conventional a priori to foundational work in general scholarship and in institutional economics, which look to the complexities of behavior and scale them toward understanding the unpredictable behaviors of modern state and market relations.[27]

At other times, new phrases have been introduced to familiarize readers with a foreign context. I have attempted to cast a critical eye on all source materials, and the careful act of weighting and arranging evidence has pressed on my work its own brand of insight and argument. For example, after observing the extraordinary lengths to which Soviet scientists went to promote economic reform with networks, I settled on the phrase *network entrepreneur* to cast a new light on the dynamics of the knowledge base in Soviet science and technological innovation. This word choice might seem misplaced because the Soviet knowledge base appears at first glance to carry none of the cultural or conceptual weight of venture capital, investment risk, and inherited responsibility for an enterprise that typically is associated with the modern English term *entrepreneur*. And yet the Soviet Internet makes a fitting case study in the global history of technology entrepreneurs, from Thomas Edison to Steve Jobs to Sergei Brin. That history has yet to be written, although when it is, it will feature an international species of actors, among them Soviets, who were prone to repeat bold slogans before proceeding by bolder failures.[28] Those who are uncomfortable applying a capitalistic term to comparable socialist practices may do well to recall that the English *entrepreneur* is already on loan from the French.

Structure

This book proceeds in five roughly chronological chapters. Chapter 1 introduces the global consolidation and spread of cybernetics as a midcentury science in search of self-governing systems from World War II to the mid-1960s. It also notes that cybernetics articulated internationally distinct scientific dialects to try to harness a range of different information systems—including biological, mechanical, and social—under one umbrella science. The term *heterarchy* is introduced as a cybernetic term for complex networks with multiple conflicting regimes of evaluation in operation at the same time. Also looked at are the mind and its neural networks (including the brain and the nervous system) as an international analogy of choice for thinking about national networks. Then the chapter examines the historical backdrop of the sequential rejection, adoption, adaptation, and mainstreaming of cybernetics in the 1950s and 1960s Soviet Union, against which the central tragedy of the remaining chapters and cast of characters unfolds.

Chapter 2 examines the emergence of economic cybernetics in the late 1950s and early 1960s as a field that was closely allied to mathematical economics and econometrics yet peculiar in its implications in the international sphere of Soviet intellectual and political influence. It also outlines and describes the basics behind the command economy and the tremendous coordination problems that the Soviet state and competing schools of orthodox, liberal, and cybernetic economists all agreed needed to be addressed and reformed in the early 1960s. A few sources of the organizational dissonance, including heterarchical networks of institutional interests, that was underlying the Soviet command economy and its state administration are also introduced.

Chapter 3 chronicles the first three aborted attempts to network the Soviet nation. The first was Anatoly Kitov's pioneering proposal in the fall of 1959 to build a nationwide computer network for civilians on preexisting military networks. The resulting show trial removed him, the first Soviet cyberneticist and a star military researcher, from the military. The second attempt was the short-lived technocratic proposal by Aleksandr Kharkevich in 1962 to build a unified communication system for standardizing and consolidating all communication signals in the Soviet Union. And the third attempt was the simultaneous proposal by N. I. Kovalev for a rational system for economic control using a nationwide web of computer networks. Brief attention is paid to the historical concurrence of cold war networks, including a caution against the cold war preoccupation to overvalue claims to being historically "first" in and outside of Soviet science.

Chapter 4 introduces the most ambitious and prominent of Soviet network projects—the All-State Automated System (OGAS)—and its primary promoter and protagonists, the cyberneticist Viktor M. Glushkov, whose stories are brought together for the first time in English. This chapter details what is known about the sweeping theoretical and practical reach of the OGAS Project between 1962 and 1969, its vision for an economy managed by network, and the institutional landscape that evolved in support of that initial project proposal in the 1960s. It also presents snapshots of both the playful work (counter)culture and informal institutional obstacles that began to preoccupy two of the most prominent research institutes for economic cyberneticists—Nikolai Fedorenko's Central Economic-Mathematical Institute (CEMI) and Viktor Glushkov's Institute of Cybernetics—in the same decade.

Chapter 5 chronicles the slow undoing of the OGAS between 1970 and 1989. Neither formally approved nor fully rejected, the OGAS Project found itself (and proposals to use computer-programmed networks to plan social and economic resources, including those by the chess grandmaster Mikhail Botvinnik) stalemated in a morass of bureaucratic barriers, mutinous ministries, and institutional infighting among a state that imagined itself as centralized but under civilian administration proved to be anything but. By the time that Mikhail Gorbachev came to power, Glushkov had died, and the political feasibility of technocratic economic reform had passed. This chapter frames how hidden social networks unraveled computer networks.

The conclusion reflects on and complicates the plain statement that is the conceit of this book—that the first global computer networks began among cooperative capitalists, not competing socialists. Borrowing from the language of Hannah Arendt, it recasts the Soviet network experience in light of other national network projects in the latter half of the twentieth century, suggesting the ways that the Soviet experience may appear uncomfortably close to our modern network situation. A few other summary observations for scholar and general-interest reader are offered in close.

1   A Global History of Cybernetics

I am thinking about something much more important than bombs. I am thinking about computers.
—John von Neumann, 1946

Cybernetics nursed early national computer network projects on both sides of the cold war. Cybernetics was a postwar systems science concerned with communication and control—and although its significance has been well documented in the history of science and technology, its implications as a carrier of early ideas about and language for computational communication have been largely neglected by communication and media scholars.[1] This chapter discusses how cybernetics became global early in the cold war, coalescing first in postwar America before diffusing to other parts of the world, especially Soviet Union after Stalin's death in 1953, as well as how Soviet cybernetics shaped the scientific regime for governing economics that eventually led to the nationwide network projects imagined in the late 1950s and early 1960s.

The term *cybernetics* evades easy definition. Today there are still more self-identified cyberneticists in the world than available definitions of the field, although the first tally is dropping as the second tally creeps slowly upward. In the English-speaking information science research environment, cybernetics failed to cohere as an institutionalized field, a fact that partially explains the inability of specialists to agree on a definition for the field. And yet the definitions are no easier in the territories of the former Soviet Union, where cybernetics did take root and still enjoys institutional recognition fifty years later. To this day, the definition puzzle holds: the postwar science remains a rich subject for critical inquiry precisely because it has escaped a clear-cut characterization.

Since the mid-1940s, cybernetics' themes of communication and control in computational biological, social, and symbolic systems have inspired and bedeviled researchers across the natural sciences, social sciences, and humanities. Accounts have identified cybernetics as a science of communication and control, a universal science, an umbrella discipline, a Manichean science, and a scientific farce founded on sloppy analogies between computers and human organisms.[2] Against its interdisciplinary backdrop of a computer-compatible formulation of communication, scores of scientists, philosophers, and policy makers advanced the midcentury computer as a tool for modeling systems and specifically for the regulation of information flows and behavior in the animal, the machine, and society. In addition to computer modeling, it gathered together preexisting concepts such as feedback loops in control systems, cooperative human-machine relations, and some foundations for the network design of digital computing. In the information sciences, it formalized midcentury mind-machine analogies that continue to animate some corners of contemporary artificial intelligence research. In the hands of polymaths such as Norbert Wiener, Warren McCulloch, and Donald MacKay, the technical and technocratic insights into a summary set of cybernetic sciences—operations research, systems theory, game theory, and information theory—presented themselves with seemingly cosmological force, delivering balance to a postwar world riven by rage.

Modern computing talk owes a fair amount to these cybernetic sciences as well. A visible contribution of cybernetics may be its consolidation and popularization of a robust vocabulary for computing, including words such as *information, control,* and *feedback*. In modern parlance, cybernetics also gave currency to the widely used and now slightly pejorative prefix *cyber* (*-bully, -café, -crime, -dating, -fraud, -law, -punk, -security, -sex, -space, -terrorism, -warfare*) as well as the phrase "in the loop." In popular culture, cybernetics also helped breathe life into the scientific fictional imagination of the cyborg—or cybernetic organism—as an ensemble of human and machinic parts, even though in practice formal cybernetics research rarely dealt with cyborg research.[3]

For the purposes of this project, cybernetics sets the scene and props up the intellectual scaffolding that is helpful for understanding the promises and problems of cold war computing initiatives and sciences, including the U.S. ARPANET, the Soviet OGAS, and OGAS's sibling network projects. In this chapter, I trace a brief global history of cybernetics, including its sources and consolidation in postwar America, its spread to other cold war climes and countries, and its adoption in post-Stalinist Soviet science.

Backlit by observations about how cybernetics, like most scientific discourse, expresses itself in variable international dialects as well as common metaphors (such as the human mind as a model information system for designing other systems), I then detail four major stages in the history of Soviet cybernetics in general and the rise of a peculiarly Soviet field—economic cybernetics—on which subsequent chapters build.

The American Consolidation of Cybernetics

Norbert Wiener, the MIT mathematician, inveterate polymath, and son of the founder of Slavic studies in America, is often credited with launching cybernetics with his 1948 book *Cybernetics, or Control and Communication in the Animal and the Machine*.[4] How much of any scientific event can be credited to one person is arguable, although we can at least credit Wiener for helping to consolidate and coin under one label a series of intellectual influences and sources. These sources were so complex and varied that perhaps his greatest accomplishment was not setting into motion a new field but synthesizing ideas from philosophy, mathematics, engineering, biology, and literary and social criticism in his masterwork. Wiener's input exceeded even his output, which was tremendous. During World War II, Wiener researched ways to integrate human gunner and analog computer agency in antiaircraft artillery fire-control systems, vaulting his wartime research on the feedback processes among humans and machines into a general science of communication and control, with the gun and gunner ensemble (the man and the antiaircraft gun cockpit) as the original image of the cyborg.[5]

To designate this new science of control and feedback mechanisms, Wiener coined the neologism *cybernetics* from the Greek word for *steersman*, which is a predecessor to the English term *governor* (there is a common consonant-vowel structure between *cybern-* and *govern—k/g* + vowel + *b/v* + *ern*). Wiener's popular masterworks ranged further still, commingling complex mathematical analysis (especially noise and stochastic processes), exposition on the promise and threat associated with automated information technology, and various speculations of social, political, and religious natures.[6] For Wiener, cybernetics was a working out of the implications of "the theory of messages" and the ways that information systems organized life, the world, and the cosmos. He found parallel structures in the communication and control systems operating in animal neural pathways, electromechanical circuits, and information flows in larger social systems.[7] The fact that his work speaks in general mathematical terms also sped his work's reception and eventual embrace by a wide range of readers, including Soviet philosopher-critics, as examined later. Wiener placed little faith in his scientific field to usher in peace—a social value disguised in his technical work on *homeostasis*, a near synonym for dynamic equilibrium that he borrowed from biology—into a world destabilized by mass violence. Nonetheless, his thesis of the 1950 second edition of his masterwork *Cybernetics* prophesied that "society can only be understood through a study of the messages and the communication facilities which belong to it; and that in the future development of these messages and communication facilities, messages between man and machines, between machines and man, and between machine and machine, are destined to play an ever-increasing part."[8]

A second strand of American cybernetic thought, led by neurophysiologist Warren McCulloch, took seriously the brain-computer analogy—that is, the now long-disputed notion that a brain can best be described as a complex information processor, transmitter, and site of memory storage.[9] McCulloch is remembered for his long white beard and contributions as the organizer of the Macy Conferences on Cybernetics, which consolidated the cybernetics movement in America. Researchers and historians of science remember his 1943 paper, coauthored with the enigmatic polymath Walter Pitts, "A Logical Calculus of the Ideas Immanent in Nervous Activity," which proposed models for neural networks in the brain that later became influential in the theory of automata, computation, and cybernetics. Their argument holds that the mind is, given certain reductions, equivalent to a Turing machine. In other words, with sufficient abstraction, it is possible to imagine the neural network in a mind as a logical circuit that is capable of carrying out any computable problem. In McCulloch's words, he sought "a theory in terms so general that the creations of God and men almost exemplify it."[10]

That "almost" packs much into its experimental epistemology. Although the conclusion that the mind functions as a computer has since been disputed and dismissed by several generations of neuroscience and cognitive science, the basic neurophysiological insights that McCulloch brought to cybernetics animated the midcentury cybernetic scene. These insights included some inspiration for the development of distributed communication networking behind the ARPANET and to this day continue to inform some contemporary artificial intelligence research. In what follows, I reintroduce his seminal but largely overlooked cybernetic notion of *heterarchy* to understand dynamic networks of competing actors.

If cybernetics in the United States sprang from the teams of researchers channeling Wiener and McCulloch, it took disciplinary shape at the

Macy Conferences on Cybernetics, a series of semiannual (1946–1947) and then annual (1948–1953) interdisciplinary gatherings chaired by War-ren McCulloch and organized by the Josiah Macy, Jr. Foundation in New York City. The Macy Conferences, as they were informally known, staked out a spacious interdisciplinary purview for cybernetic research.[11] In addi-tion to McCulloch, who directed the conferences, a few noted participants included Wiener himself, the mathematician and game theorist John von Neumann, leading anthropologist Margaret Mead and her then husband Gregory Bateson, founding information theorist and engineer Claude Shan-non, sociologist-statistician and communication theorist Paul Lazarsfeld, psychologist and computer scientist J.C.R. Licklider, as well as influential psychiatrists, psychoanalysts, and philosophers such as Kurt Lewin, F.S.C. Northrop, Molly Harrower, and Lawrence Kubie, among others. Relying on mathematical and formal definitions of communication, participants rendered permeable the boundaries that distinguished humans, machines, and animals as information systems. The language of cybernetic and infor-matic analysis—including terms such as *encoding, decoding, signal, feedback, entropy, equilibrium, information, communication, control*—sustained the anal-ogies that bound together ontologically distinct physical phenomena.[12]

The "invisible college" constituted by the Macy Conferences proved immensely influential:[13] Von Neumann pioneered much of the digital architecture for the computer as well as cold war game theory;[14] Shannon founded American information theory; Bateson facilitated the adaptation of cybernetics in anthropology and the American counterculture;[15] Lazars-feld fashioned much of postwar American mass communication research;[16] and of special note here, Licklider went on to pioneer and manage the U.S. ARPANET (predecessor to the Internet) and its founding vision of human-computer interaction. The effects of World War II on the global research community shaped both the number of international participants in the group (for example, von Neumann was a Hungarian émigré and Lazars-feld was Viennese) as well as the distinctly American approach that the Macy Conferences represented as a trading zone between private philan-thropic institutions (the Macy Foundation) and academics with strong ties to U.S. military research (including von Neumann, Wiener, Bateson, and many others).[17] Cybernetics emerged as a discipline that consolidated dis-tinctly international sources of inspiration in a distinctly postwar American setting.

The principles to emerge out of the Macy Conferences were many and far from consensual. "Our consensus has never been unanimous," McCulloch quipped in a summary of the proceedings: "Even had it been so, I see no reason why God should have agreed with us." Nonetheless, a few remarks help sketch out its conceptual pliability for later international interpretation. The first methodological hallmark of cybernetics is that it is not one thing but that its key concepts, especially human-machine interaction and feedback, outline a kind of vocabulary for working analogically across different systems—computational, mechanical, neurological, organic, social— that rendered its vocabulary fecund for other sibling fields embedded in U.S. military-industrial research.[18]

Take, for example, the contemporary fields of information theory and game theory. Mainstream American information theory, following Bell Labs engineer Claude E. Shannon's 1948 mathematical theory of communication, concentrates on the efficient and reliable measurement and transmission of data.[19] Perhaps its central seminal contribution is the theorizing of a statistical framework for understanding all data transmissions. All communication messages became a question of probabilities and stochastic analysis, and the term *information* abandoned its ordinary meaning of relevant facts and took on a new definition as a technical measure of the likelihood that a message contains something ordered or surprising. Such insights sped the theoretical development of computational communication systems, although Shannon's theories were not widely applied until the advent of affordable personal computers in the 1970s.

Von Neumann's game theory (still influential in contemporary economics, business, and policy) developed formal models for human behavior based on strategic and rational decision-making processes.[20] By presuming that the players in its games are rational actors seeking to make strategic decisions, game theory formalized approaches to mathematically describing, modeling, and proscribing the optimal behaviors in both competitive and cooperative multiplayer interactions that came to characterize the cold war as a whole.[21]

The founders of these fields disagreed about the limits and relationships between the three fields. Shannon insisted on keeping the technical principles of information theory separate from the more sweeping scope of cybernetics, Von Neumann did not rigorously distinguish between the three, and Wiener defended his grouping of the other two research fields under the cybernetics umbrella, even as (especially after mid-1950s) many information theorists and game theorists objected to any conflation of these fields.[22] All three fields presented overlapping rational and generalized models of communication, or a "theory of messages" fit for application, even though no one—not even the founders—knew the exact limits of these computation communication sciences. Shannon did not accept the label of cybernetics, and he also did not accept the label others had given to his own "information theory," preferring to the end of his life his original emphasis on "mathematical theory of *communication*." Each of these sciences sought to theorize the technical means by which communication could be controlled. The cybernetic sciences, especially but not exclusively in the Soviet case, emerge as a communication science in search of self-governing systems.

Although it has never been clear (perhaps even to cyberneticists) what cyberneticists could do exactly, it also never has been obvious what cybernetics *could not* do (perhaps even the definition of cybernetics is self-governing). For example, in 1943, Wiener and his coauthors succeeded in springing "feed-back"—a once obscure term on loan from control engineering and reclaimed in his antiaircraft research—into an umbrella concept that was fit for understanding any type of purposeful behavior, where the behavior of humans, animals, and machines is understood as "any change of an entity with respect to its surroundings."[23] At this philosophical height, feedback loops proved a generalizable tool that could stabilize all kinds of unsteady systems: feedback offers a process whereby information that leaves a system is brought back into the system with the intention of influencing that system's future behavior.

Feedback comes in at least two kinds—positive and negative. When positive, feedback amplifies a signal cyclically, much like a microphone that is set too close to a loudspeaker will cause painful audio feedback as the signal loops out of control. Negative feedback by contrast can serve as a stabilizing agent, an internal check or correction on a system seeking balance in an unstable environment. By working with feedback loops in communication systems, cybernetics sought a revolution in recognizing and operationalizing the nonlinear, self-recursive processes that abound in nature and technology. Whatever cybernetics is, it is not a straightforward worldview of Newtonian physics, Cartesian grids, Euclidean geometry, Aristotelian cause and effect, and arithmetic. Rather, cybernetics espouses a mathematical worldview that helps us understand the midcentury struggle to balance atop the tectonic shift in science toward pre-postmodern concepts such as quantum physics, curvilinear grids, non-Euclidian geometry, cyclical causalities, self-similar fractals, and modern probability theory.

The interpretive purchase of Wiener's cybernetics rests not on its clarity but on its synthetizing search for system self-regulation in the face of a topsy-turvy postwar world. That is, the basic cybernetic approach seeks to harness to the logical power of computing a wide range of scientific problems with circularities and feedback loops. In this search for a balance between the incongruities of material behavior and the sharp logic of computing, feedback—even more than the cybernetic watchwords *information*, *control*, and *equilibrium*—emerges as a clean concept for attempting to domesticate all kinds of unruly communication systems. In 1943, Warren McCulloch introduced the companion, although largely neglected, notion of *heterarchy*, which serves as a useful lens for focusing scrutiny on the Soviet case. This cybernetic concept helps describe some of the sources of conflict that beset Soviet cybernetic attempts to network their command economy.

Let us set up this argument with a glimpse into institutional networks of actors that are based on neither flat market nor hierarchical states but on this third or middle way of *heterarchy*. The cold war organizational tropes for self-regulation break down along that spectrum of economic order that conventionally opposes market and hierarchy. In this view, the *market* is understood as a flattened space for free interaction and efficient possibility discovery among varied economic actors, and the *hierarchy* is understood as a well-ordered, top-down pyramid of superiors over subordinates that is well suited for completing long-term and complex tasks. Etymologically, the English *market* is by far the newcomer of the two and can be traced back to the mid-thirteenth-century Italian term for a "public building or space for trading, buying, and selling." The term *market economy* is first noted in English only in 1948, centuries after the early modern capitalist revolution that gave it fame and that has since enjoyed a privileged if often misunderstood position in the Western vocabulary of modern politics, economics, and society. One reason for justifying the Pareto efficiency of the market rests on the transitivity of human preferences. For the market to be the ideal organizational mode, some economists assume that rational actors will rank the order of their preferences linearly: if rational actors prefer option A over B as well as option B over C, they also will prefer option A over C. Yet this view of the market has been challenged in recent decades. Markets hide transaction costs and information asymmetries. Behavioral economists have demonstrated how under a number of conditions (such as fear, regret, the threat of loss, cognitive dissonance, or peer pressure) the rational *homo economicus* is a fiction: a person may prefer apples to bananas, bananas to cantaloupes, and cantaloupes to apples, and there is no guarantee that there exists a rational solution to voting systems or daily choices involving three or more actors.[24]

By contrast, the concept of *hierarchy* (from the Greek term ἱεραρχία, "rule by priests") reaches back fifteen centuries to religious roots. As sociologist of economics David Stark shows, the term was first used by a Christian medieval theologian who is known today as the Pseudo-Dionysius the Areopagite.

He published in the late fifth century under the pseudonym Dionysius the Areopagite, a name first attributed to a first-century convert of Paul and the first bishop of Athens. The fifth-century Christian mystic theologian describes two far-reaching hierarchies in his *Heavenly and Ecclesiastical Hierarchies*: the first, which includes nine levels of celestial beings (extending from the supreme Godhead to the angels, who were just above humans), serves as a symmetrical reflection for the second, which includes nine levels of church leadership (seen in the current nine-tier Catholic ecclesiastical hierarchy descending from pope to bishop).[25] The concept of hierarchy has abounded in Western thought ever since—in the nine levels in Dante's *Inferno*; the organizational design of countless church, military, governmental organizations; and the conceptual imprint of information classification systems, computer sciences, mathematics, and categorical thought. These are all scalable approaches to bringing order in the modern world. Perhaps the strongest example of hierarchy and socialism in modern America is also its greatest bastion of patriotism—the U.S. armed forces, whose command-and-control silos deliver social services and benefits to its members.

The logic of hierarchy has faced many challenges. Most modern critical thought—epistemologists William James and Michel Foucault, critical theorists and feminists, Marxists and free-market theorists, liberation theorists and theologians on the radical left and right, digital media theorists and others—is organized *against* hierarchy.[26] Even though the cold war ideological division over planned and free-market economies preoccupies fewer social scientists today, modern organizational power and its resistance still organize along the coordinates of hierarchy and open system.

The cybernetic concept of *heterarchy* offers a third way and an alternative model between market and hierarchy that helps make sense of the Soviet cyberneticists and informs later network analysis of how Soviet cyberneticists tried to build computer networks to match the institutional networks running the command economy. In 1945, just before McCulloch took stewardship over the Macy Conferences, he published a five-page essay, "A Heterarchy of Values Determined by the Topology of Nervous Nets," in the *Bulletin of Mathematical Biophysics* that coined the term *heterarchy* and established how even the simplest systems can be subject to multiple competing regimes of evaluation.[27]

Heterarchies are neither ordered nor disordered but instead are ordered complexly in ways that cannot be described linearly. McCulloch takes as his simplest example a network of three neurons arranged into a hierarchy of transitive connections from neuron A to neuron B and from neuron B to neuron C in which there are no "diallels" or "cross-overs." His description references the hierarchy in "the sacerdotal structure of the Church" in which "the many ends are ordered by the right of each to inhibit all inferiors." He then contrasts a hierarchical network with an intransitive neural network in which a crossover is introduced between neurons C and A. In this case, to model the network one needs to "map the network not on a plane, but on a three-dimensional Taurus (a donut-shaped topological space)." Instead of imagining such a network arrangement as inferior or inconsistent, he observes that "circularities in preference actually demonstrate consistency of a higher order than had been dreamed of in our philosophy. An organism possessed of this nervous system—six neurons—is sufficiently endowed to be unpredictable from any theory founded on a scale of values. It has a heterarchy of values, and is thus internectively too rich to submit to a *summum bonum*."[28]

With the concept of neural *heterarchy*, McCulloch introduces the multidimensional possibilities for complex systems that cannot be mapped onto two-dimensional logics of either flat markets or tall hierarchies. This concept has since proven helpful in cybernetic-compatible research far beyond brain research, including self-organization, feedback loops, automata theory, and non-Turing and non-Euclidean computing for thinking about the superabundance of actual complex networked relations and also about the limits of traditional tools for accounting for these relations. As detailed later, similar cybernetic notions introduced both the terms and the network tools for describing and managing the heterarchical tensions at the heart of the Soviet command economy.

Cybernetics beyond the Cold War Superpowers

Between 1948 and the mid-1950s, cybernetics also enjoyed reception and development in a number of countries outside of the cold war superpower axis. For the purposes of this section, I focus on the postwar reception of cybernetics in England, France, and Chile, although the point of the section supersedes comparative local or national histories. The conditions of modern countries after World War II and during the cold war were ripe for an umbrella science of self-governance. Many scientists worldwide were rushing to find ways to stabilize and regulate the consequences of a torrent of new and disruptive technologies—and cybernetics modeled a technical mindset for how to grapple with and control the consequences of technology itself. The 1950s saw a dizzying number of potentially revolutionary technologies become popular—atomic and hydrogen bombs, nuclear power plants, *Sputnik*, the double helix, passenger jets, dishwashers, polio vaccines, the lobotomy (invented in the 1930s), television, and transistor radios—and other trends, such as rock & roll and suburban housing developments. The disruptive influences of modern science and technology continued to be felt in the 1960s as quarks, lasers, *Apollo*, nylon, Pampers, the pill, LSD, napalm, DDT, mutually assured destruction, and the ARPANET entered the world stage. The most disruptive and destructive of all was the development of computers around the work of John von Neumann at the Institute of Advanced Studies at Princeton to study and control the effects of nuclear bombs.[29] The technocratic promise of the computer seemed to promise both delivery and destruction. If computers could help civilize the terrible and awesome power of the atom bomb, thought the scientists of the day, then perhaps it might help stabilize lesser disruptions of modern science and technology. If not, what terrible consequences would follow? Or as von Neumann asked, taking the pulse of the moment in 1955: "Can We Survive Technology?"[30]

Von Neumann's question especially animated those who were engaged in the nuclear cold war. In postwar France, United Kingdom, and Chile, the potential of the computer to civilize awesome powers generated a "technology" of cybernetic interest in the 1950s and 1960s that was sometimes more disruptive than the atomic bomb that troubled von Neumann. It was the human mind imagined as an embodied machine. Might cybernetics and its heir in cognitive science, midcentury scientists wondered, crack the human mind and in turn spark new insights into how that most creative of technologies might be modeled elsewhere?[31]

In France, the intellectual contributions of cybernetics began with more analogies to politics than to the parietal lobe. Cybernetics had an early start and a long afterlife in postwar France for several reasons. The public debate about cybernetics turned the science into a bit of a political football between communist and anticommunist debates in postwar France; the local intellectuals helped ascribe a long French intellectual tradition to cybernetics, which softened its reception; and Norbert Wiener visited France repeatedly and promoted his science vocabulary in person. The imprint of cybernetics can still be seen in subsequent generations of French theorists.

These postwar happenings are described briefly below. In 1947, the year before he published *Cybernetics* with the MIT Press, Wiener attended Szolem Mandelbrot's congress on harmonic analysis in Nancy, France, which resulted in a French book contract for the book that, while initially resisted by the MIT Press, sold a sensational 21,000 copies over three reprints in six months after its release in 1948. Three years later, in 1951, at the invitation of Benoit Mandelbrot, the founder of fractals and Szolem's nephew, Wiener returned to lecture at Collège de France. Between 1947 and 1952, a flurry of press coverage and public controversy sprung up between two camps of anticybernetic communists and anticommunist cyberneticists.[32] (Jacques Lacan, who served in the French army, may very well have been among the anticommunists and early cyberneticists at the time.) These debates over the future of the governance of the French state were fueled by a slow and painful postwar recovery, with widespread poverty aiding popular communist and pro-Soviet sentiment. However, after Paul Ramadier's socialist party voted to accept the American Marshall Plan during the international Paris meeting in 1947, anticybernetic communists slowly fell out of public favor and with it, the debate about cybernetics. Similar to the initial Soviet rejection of cybernetics, the initial public reaction to cybernetics appears less about its science than about its status as an American import.[33]

At the same time, the public and implicitly pro-American defense of cybernetics in the French press also helped reclaim this foreign science as the heir to a distinctly French intellectual tradition that included the rational mind-body concerns of René Descartes' and Denis Diderot's rational encyclopedia, the physicist André-Marie Ampère's coining of the term *cybernétique* as a political science of peaceful governance in 1834, and the structural linguistics of Ferdinand Saussure. Although I currently know of no obvious direct connection between French cybernetics and the Minitel network that developed between 1980 and 1989, the situation nonetheless points to a generative and transnational intellectual exchange about the scientific self-governance of a nation. As recent interpreters have argued, many leading lights of postmodern French theory trace some of their basic insights to French postwar cybernetic sciences. These include Claude Lévi-Strauss's treatment of language as a technologically ordered series (after meeting with Macy Conference attendee Roman Jakobson in Paris in 1950); Jacques Lacan's turning to mathematical concepts; Roland Barthes's turn to schematic accounts of communication; Gilles Deleuze's abandonment of meaning, with Claude Shannon's information theory in hand; Felix Guattari's, Michel Foucault's, and other French theorists' experimentation with terms such as *encoding, decoding, information,* and *communication.*[34] Postmodern French theory owes a deep debt to postwar information theory and the cybernetic sciences.

In England, cybernetics took on a different character in the form of the Ratio Club, a small but potent gathering of British cybernetic figures who gathered regularly in the basement of the National Hospital for Nervous Diseases in London from 1949 through 1955. Notable figures include the

computing pioneer Alan Turing, his Bletchey Park colleague and cryptog-
rapher mathematician I. J. Good, neuropsychologist Donald MacKay, and
astrophysicist Tommy Gold. The historian of science Andrew Pickering
chronicles the lives and work of six active and largely forgotten Britons who
were preoccupied with what the brain does—neurologist W. Grey Walter,
psychiatrist W. Ross Ashby, anthropologist, psychiatrist, and Macy Conference attendee Gregory Bateson, radical antipsychiatrist R. D. Laing, psychologist Gordon Pask, and management cyberneticist Stafford Beer (who
also features prominently in the Chilean cybernetic situation described
below). Interdisciplinary discussions ranged widely across themes such as
information theory, probability, pattern recognition, artifacts that act (such
as William Ross Ashby's homeostat and W. Grey Walter's robotic tortoises),
and philosophy. Among their guests over the years were at least two Americans who later played roles in the development of the ARPANET—J.C.R.
Licklider and Warren McCulloch.[35] According to Pickering, what each of
these pioneering cyberneticists held in common was an interest in the
brain as a machine that acts, not thinks—or communication systems that
perform, not cogitate.[36] Cybernetics took root on its own terms in Britain—
not the postmodern theory of France but what Pickering calls the "non-
modern" performances of neurological structures.

    Chile and to a lesser extend Argentina also experienced the influx of
cybernetic ideas that ended up framing the debate about national networks.
In 1959, as a graduate student at Harvard, the Chilean biologist Humberto
Maturana coauthored an important paper, "What the Frog's Eye Tells the
Frog's Brain" with lead author Jerome Lettvin, Warren McCulloch, and
Walter Pitts. In the early 1970s, Maturana and his student Francisco Valera
secured their part in what has been called the "second wave" of cybernetics together with the editor of the Macy Conferences proceedings Heinz
von Forester and Gordon Pask, among others, with their contribution of
the idea of *autopoiesis*—a system that generates, maintains, and reproduces
itself (such as a biological cell). The idea found resonance with the work on
the Chilean socialist economy led by the British management cyberneticist
Stafford Beer during the political rule of Salvador Allende between 1971
and the military coup in 1973.

    During this period, Project Cybersyn took place, and it was perhaps the
most prominent experiment in developing a national network intended
for managing the socialist economy. As historian of science Eden Medina
has recently revealed, the British management cyberneticist Stafford Beer
served as principal architect for the rapid design, development, and partial
deployment of this nationwide network of telex machines connected to

a central mainframe computer. Beer, working with then finance minister and engineer Fernando Flores, imported and adapted his (British) emphasis on the brain as a model for managing organizations as published in the 1972 book *The Brain of the Firm*.[37] His overtly cybernetic idea of a viable system—a system that is designed to survive by adapting to its changing environment—took root in the design of the Project Cybersyn network and was reflected the political ideals of Allende's democratically elected socialism and the autonomy of the workers. Despite limited success in rerouting goods during a 1972 strike of truck drivers, the Cybersyn network, including its futuristic central operations room, were scrapped in the military coup of General Augusto Pinochet in 1973. In this Chilean case and perhaps in the larger Latin American scene, cybernetics dovetailed with a strong emphasis on embodied philosophy of mind.[38]

Before turning to the Soviet reception and translation of cybernetics, let us look briefly at the eastern European sources of the cybernetic tradition, some of which precede its consolidation in U.S. military research and the postwar Macy Conferences. The list of precybernetic promoters includes several notable figures. Aleksandr Bogdanov—old Bolshevik revolutionary, right-hand man to Vladimir Lenin, and philosopher—developed a wholesale theory that analogized between society and political economy, which he published in 1913 as *Tektology: A Universal Organizational Science*, a proto-cybernetics minus the mathematics, whose work Wiener may have seen in translation in the 1920s or 1930s.[39] Stefan Odobleja was a largely ignored Romanian whose pre–World War II work prefaced cybernetic thought.[40] John von Neumann, the architect of the modern computer, a founding game theorist, and a Macy Conference participant, was a Hungarian émigré. Szolem Mandelbrojt, a Jewish Polish scientist and uncle of fractal founder Benoit Mandelbrot, organized Wiener's collaboration on harmonic analysis and Brownian motion in 1950 in Nancy, France. Roman Jakobson, the aforementioned structural linguist, a collaborator in the Macy Conferences, and a Russian émigré, held the chair in Slavic studies at Harvard founded by Norbert Wiener's father. And finally, Wiener's own domineering and brilliant father, Leo Wiener, was a self-made polymath, the preeminent translator of Tolstoy into English in the twentieth-century, the founder of Slavic studies in America, an émigré from a Belarusian shtetl, and like his son, a humanist committed to uncovering methods for nearly universal communication.[41]

Although summarizing the intellectual and international sources for the consolidation of cybernetics as a midcentury science for self-governing systems is beyond the scope of this project, the following statement is probably not too far of a stretch. In each of the case studies examined here—Warren McCulloch's heterarchical neural networks, the French evolution of information theoretic and its turn to postmodern theory, the British Ratio Club's emphasis on performative models and agents, the Chilean building of a socialist national economic network after a model of the nation as an organized firm, and sundry competing eastern European forces—the midcentury cybernetic sciences are expressed in the local dialects of an intellectual milieu and share with cognitive science an impulse to think with the model of the mind. To a different effect, cyberneticists have been constructing system analogs to understanding the mind and using such mind models as analogs for reenvisioning new social, technological, and organic worlds. The fascination with the mind is not new to cybernetics. The millennia-long preoccupation with the inner workings of the mind, as one neuroscientist quipped, may be little more than our own brain's conceit about itself.[42]

Soviet Cybernetics

With the first Soviet test of the atomic bomb in 1949, the cold war conflict between capitalism and socialism slipped into the nuclear age. Soviet scientists, philosopher-critics, and journalists redoubled their search for real threats, as well as exciting possibilities, in the rapidly developing sphere of science and technology, including rumors about a new American field called *cybernetics*. Between 1947 (the year Norbert Wiener coined the term *cybernetics* at a Macy Conference in New York) and 1953 (the year after Joseph Stalin died), the state of Stalinist science, having proven itself as essential to winning the war, enjoyed a complicated improvement in social status, better funding, and uneven intellectual autonomy.[43] The Soviet Union stood out as a state that was committed to groundbreaking science.[44]

At the same time, certain fields of science, especially genetics in the wake of the Lysenko debates, experienced acute pressures and censorship.[45] And although cybernetics was not outright repressed during Stalin's rule, it was widely ridiculed in the press and did not flourish until after his death. The remainder of this chapter shows that even though post-Stalinist cybernetics seemed poised to remake the Soviet Union as an information society, the history of Soviet cybernetics, especially during the period of its rehabilitation and adoption, slouches in significant ways toward the normal patterns of Soviet history. In four overlapping sections below, I show that Soviet scientific discourse rejected, rehabilitated, adopted, and adapted cybernetics for historically expedient and changing purposes.

The Stalinist Campaign against Cybernetics: A "Normal" Pseudo-Science

Not all was rosy at the start. Amid abundant American accolades following the publication of Wiener's *Cybernetics, or Control and Communication in Animal and Machine* in 1948, the Soviet press poured on insults. In 1950, at the same time that the American *Saturday Review of Literature* was triumphantly proclaiming that it was "impossible for anyone seriously interested in our civilization to ignore [Wiener's *Cybernetics*]. This is a 'must' book for those in every branch of science," the leading literary Soviet journal *Literaturnaya gazeta* was calling Wiener one of those "charlatans and obscurantists, whom capitalists substitute for genuine scientists."[46] In a 1950 article titled after the computing machine developed by Howard Aiken, "Mark III, a Calculator," Soviet journalist Boris Agapov ridiculed the sensationalist American press for its exultations about the coming era of "thinking machines," styling Norbert Wiener as an unknown figure "except for the fact that he is already old (although still brisk), very fleshy, and smokes cigars." Commenting on a *Time* magazine cover of a computer dressed in a military uniform, Agapov continued, "it becomes immediately clear in whose service is employed this 'hero of the week,' this sensational machine, as well as all of science and technology in America!"[47] After Agapov's 1950 article, Wiener's *Cybernetics* was officially removed from regular circulation in Soviet research libraries; apparently only secret military libraries retained copies into the early 1950s.[48]

In 1951, a public campaign in the Soviet Union called the computer hype in the United States a "giant-scale campaign of mass delusion of ordinary people." The 1951 volume *Against the Philosophical Henchmen of American-English Imperialism* categorized cybernetics as part of a worrying fashion around "semantic idealism" and dubbed cyberneticists "semanticists-cannibals" for their recursive logics, especially self-informing feedback loops. In addition to American cyberneticist Norbert Wiener, the volume identified those belonging to the group of "semantic obscurantists" as including logician-pacifist Bertrand Russell, his Cambridge colleague Alfred North Whitehead, and Vienna Circle logical positivist Rudolf Carnap. Positivism, semiotics, and mathematical logic all appeared guilty of the cardinal cognitivist belief that "thinking was nothing else than operations with signs."[49] In 1952, *Literaturnaya gazeta* ran an article called "Cybernetics: A 'Science' of Obscurantists," which cleared the way for a deluge of popular titles: "Cybernetics: An American Pseudo-Science," "The Science of Modern Slaveholders," "Cybernetics: A Pseudo-Science of Machines, Animals, Men and Society," and so on.[50]

In 1953, an author who wrote under the pseudonym "Materialist" published the infamous article "Whom Does Cybernetics Serve?" in a leading journal for ideological and intellectual battles, *Questions of Philosophy*. "Materialist" waxes poetic in his rebuke:

the theory of cybernetics, trying to extend the principles of modern computing machines to a variety of natural and social phenomena without due regard for their qualitative peculiarities, is mechanicism turning into idealism. It is a sterile flower of the tree of knowledge arriving as a result of a one-sided and exaggerated blowing up of a particular trait of epistemology.[51]

Later in the article, Materialist contends that "in the depth of their despair, [those in the capitalist world] resort to the help of pseudo-sciences giving them some shadow of expectation to lengthen their survival."[52] With somewhat less vitriol, in 1954, the fourth edition of the *Concise Dictionary of Philosophy* cast cybernetics as a slightly ridiculous, although still harmful anti-Marxist "reactionary pseudo-science." The entry reads:

Cybernetics: a reactionary pseudo-science that appeared in the U.S.A. after World War II and also spread through other capitalist countries. Cybernetics clearly reflects one of the basic features of the bourgeois worldview—its inhumanity, striving to transform workers into an extension of the machine, into a tool of production, and an instrument of war. At the same time, for cybernetics an imperialistic utopia is characteristic—replacing living, thinking man, fighting for his interests, by a machine, both in industry and in war. The instigators of a new world war use cybernetics in their dirty, practical affairs.[53]

The campaign continued in the popular and scholarly press more or less unabated through the 1950s, although the first public rehabilitation efforts, noted below, began in earnest as early as 1955.

The list of epithets reserved for cybernetics by the Soviet press should be put into perspective. The campaign against cybernetics, however mean-spirited and aggressive, appears far from the most vicious of campaigns that were organized by Soviet journalists and public commentators against American thought. Stalin, who was known to read widely across the scientific fields, seems to have known little to nothing about cybernetics; his fury against it appeared independent of "any essential features of cybernetics itself," according to Gerovitch.[54] Without any direct evidence of Stalin's involvement in the campaign against cybernetics, we can speculate that Stalin likely reviled cybernetics for the same reasons that he hated all imperialist "pseudo-sciences": ideological opposition was necessary to fuel and power his monumental state building and modernization projects. The campaign against cybernetics, which came in the wake of Stalin's personal affront against classical genetics, appeared more or less a "farce" to some philosopher-critics. These same philosopher-critics, according to information theorist Ilia Novik, "berated cybernetics with certain … indifference and even fatigue." In the late 1940s and early 1950s, as cybernetics was sweeping the United States, France, the United Kingdom, Chile, and other countries with the enthralling possibilities of self-organizing human-machine ensembles and predictive negative feedback loops, "cybernetics" in the Soviet Union had, to crib Novik's phrase, "emerged as a normal pseudo-science."[55]

The anti-American Soviet campaign against cybernetics was only one among a range of operations that were meant to repress the Soviet knowledge base, including but not limited to Stalinist science. A few other examples include the rise of Trofim Lysenko in Soviet biology, whose program on the heritability of acquired characteristics ousted the study of Mendeleev and classical genetics; the condemnation of Linus Pauling's structural resonance theory by Soviet chemists in 1951; the banning of Soviet Lev Vygotsky's work, now recognized as a foundation of cultural-historical psychology; the forestalling of structural linguistics pioneered by Ferdinand Saussure, Nikolai Trubetzkoi, and Roman Jakobson; and the excoriation of Albert Einstein's theories of general and special relativity, quantum mechanics, and Werner Heisenberg's principles of indeterminacy as distortions and corruptions of the true (that is, Marxist) objective and material nature of the universe.[56] In light of these and other examples, the public campaigns against cybernetics strike the contemporary observer as far from masterfully orchestrated or even normal in their regularity. The ground warfare of ideological critique was messy, full of ritual elements, political posturing, and routine debates. Not only did the enterprise of Soviet cybernetics prove to be diverse, but the anticybernetic campaigns that preceded it varied richly.

There was nothing particularly anticybernetic about the early anticybernetic campaigns. Rather, the early opposition to the science appears overwhelmingly anti-American in motivation. In the decade that followed, Soviet cybernetics transformed into an apparent harbinger of social reform and later into a normal Soviet science. Even the Soviet ideological resistance to cybernetics appears normal from the beginning.

The Post-Stalinist Rehabilitation of Cybernetics, 1954 to 1959

Natural Science will in time incorporate into itself the science of man, just as the science of man will incorporate into itself natural science: there will be *one* science.

—Karl Marx, *Economic and Philosophic Manuscripts of 1844*

Stalin's death in March 1953 made possible a watershed shift in public discourse in favor of Soviet cybernetics and gave root to the promise of cybernetic-led structural reform of the Soviet Union—and especially the promise of a new kind of self-governance in the wake of Stalin's bloody rule. After he seized power from his rivals in 1955, Nikita Khrushchev titled himself first secretary, not general secretary as Stalin had, in an effort to signal a clean break from the past and the launching of a new post-Stalinist era. Typically, the only thing remembered about the Twentieth Congress of the Communist Party of the Soviet Union, is Khrushchev's 1956 "secret speech," which he delivered to a carefully selected crowd and in which he became the first Soviet authority figure to denounce Stalin's crimes and the now infamous "cult of personality." The speech inaugurated the Khrushchev thaw, a period known for the easing of censorship and political repression and the partial de-Stalinization of Soviet policy, international relations, and society. These public revelations, combined with a sagging Soviet economy, compelled even those least likely to decry the terrible reality of Stalin's terror to admit that, in Khrushchev's terms, "serious excesses" and "abuses" had been committed.[57]

As part of this sweeping technical reform, the new first secretary also called for an ideological reappraisal of Marxism-Leninism:

In this connection we will be forced to do much work in order to examine critically from the Marxist-Leninist viewpoint and to correct the widespread erroneous views connected with the cult of personality in the sphere of history, philosophy, economy, and other sciences, as well as in literature and the fine arts. It is especially necessary that in the immediate future we compile a serious textbook of the history of our Party, which will be edited with scientific Marxist objectivity.[58]

By 1959, Stalin's *Short History of the Communist Party of the Soviet Union*, once characterized as "the catechism of Communism," had been officially deemed full of errors and withdrawn under Khrushchev. It was replaced in 1961 by the 900-page *Fundamentals of Marxism-Leninism*.[59]

The time between Stalin's death and cybernetics' entrance into the favor of the press and Soviet public discourse on science was not great. In fact, in the same 1956 Congress that he gave his "secret speech," Khrushchev also promoted cybernetic-friendly principles for automating the Soviet economy: "The automation of machines and operations," he declared, "must be extended to the automation of factory departments and technological processes and to the construction of fully automatic plans."[60] With the passing of Stalin, cybernetics entered Soviet technical, scientific, and political discourse at a time that was particularly primed for reform.

Although Soviet science enjoyed reforms and looser ideological constraints under Khrushchev, Soviet science may have accomplished more under the fist of Stalin than it did under the loose umbrella of cybernetics. Under Stalin, Soviet physicists and chemists pioneered work for which chemist Nikolai Semyonov, physicist Igor Tamm, economist Leonid Kantorovich, and physicist Pyotr Kapitza received Nobel Prizes decades later. Other Soviet scientists—including Igor Kurchatov, Lev Landau, Yakov Frenkel, and Andrei Sakharov, and other world-renowned figures—also developed atomic and thermonuclear bombs, a lynchpin in Stalin's rapid and forceful industrialization of the remnants of the Russian empire from a backwater country into a global super power in the period of a few decades. Many Soviet scientists successfully employed dialectical materialism as a genuine source of inspiration, not a forced ideology, in their scientific work. The reality that the health of science depended more on funding than it did on freedom also sobers reflection on the contemporary state of science and public attitudes about it.[61]

Soviet cybernetics arrived at a time that was well suited for leveraging a post-Stalinist revision of scientific Marxist objectivity. It introduced its mind-machine analogies in a light that was friendly to Ivan Pavlov's celebrated notion of "conditioned reflexes" in psychology, which were based on the reflex-response analogy of a telephone electrical switchboard, the reactions of which depended on the programmable configuration of wires. Both Pavlov and, two generations later, cyberneticists worldwide imagined the mind as neural networks and electronic processors, a seminal metaphor for what philosopher Pierre Dupuy dubbed the "mechanization of mind" powering the subsequent rise of cognitive science.[62]

Soviet cybernetics also found the support of several world-famous mathematicians, which was a field in which the Soviets were internationally recognized. Figures including Andrei Kolmogorov, Sergei Sobolev, Aleksei Lyapunov, and Andrei Markov Jr., came together, despite significant differences, to form an early core of Soviet cybernetic mathematicians who were committed to advancing this new metamathematical science as a single science for Soviet thought. And just as cybernetics was mobilizing its intellectual defenses, it also found institutional fortification in the creation of Akademgorodok, a new "scientific township at Novosibirsk" in Siberia. Created in the spring of 1957, this city of science (formally part of the city of Novosibirsk) proved a refuge of privilege and relative intellectual freedom for over 65,000 Soviet scientists, including Aleksei Lyapunov, a pioneering cyberneticist.[63]

Before the Soviet scientific mainstream could adopt cybernetics, the attendant scholarly communities had to be prepared for an about-face in the official Soviet attitude toward an American-born discipline. The first sign of this turnaround came not from Moscow but from a neighbor in the near abroad: in 1954 in Warsaw, six "Dialogues on Cybernetics" surfaced, and they approached cybernetics in a critical dialectical tone that was serious enough to signify that the topic deserved real discussion.[64] In the meantime, three mathematicians and an unlikely philosopher-critic closer to Moscow set off on a mission to remake Soviet cybernetics from the inside out.

The First Soviet Cyberneticists: Kitov, Lyapunov, Sobolev

In 1955, two Russian-language articles appeared in the same issue of the Soviet journal *Voprosi Filosofii (Problems of Philosophy)*, where "Materialist" and others had railed against cybernetics in 1953. This signaled a watershed change in the official attitude toward cybernetics. A closer look at these two articles sheds light on this reversal. Sergei Sobolev, Aleksei Lyapunov, and Anatoly Kitov coauthored the article titled "The Main Features of Cybernetics" and began the process of rehabilitating cybernetics from positions of relative authority in the Moscow military-academy complex. Although Kitov was the youngest and the least influential of the three mathematician coauthors, he also appears to have been the first Soviet cyberneticist.

A Soviet colonel engineer, Anatoly Kitov discovered in 1952 the single copy of Wiener's *Cybernetics* in a secret library of the Special Construction Bureau—SKB-245—at the Ministry of Machine and Instrument Building. Kitov had been sent there to research possible military applications for computers after graduating in 1950 from the military academy where Lyapunov taught with a gold medal, the highest award in the Soviet education system. After reading Wiener's *Cybernetics*, Kitov began to consider that cybernetics was, in his words, "not a bourgeois pseudo-science, as official publications considered it at the time, but the opposite—a serious, important science."[65]

After digesting *Cybernetics*, Kitov turned to share his newfound enthusiasm for the science with his former instructor, Aleksei Lyapunov. Lyapunov, who later was known as "the father of Soviet cybernetics," was a wide-ranging and luminous mathematician who taught at the Military Artillery Engineering Academy and in the department of computational mathematics at Moscow University. Recognized by biologists, geophysicists, and philosophers alike, Lyapunov took, according to Soviet historian of science M. G. Haase-Rapoport, an "integrating, non-dividing approach in natural science," which "became the rich soil [for] the sprout of cybernetic ideas."[66] Having heard his case, Lyapunov in turn encouraged Kitov to write an article explaining the essence of cybernetics, promising to coauthor it with him. Holed up in the secret military research library, Kitov wrote a draft for the article, after which Lyapunov recommended inviting as coauthor Sergei Sobolev, then chair of the department of computational mathematics at Moscow University. Sobolev also played a legitimizing role as deputy director of the Institute of Atomic Energy, in effect the mathematician with a hand on the atomic bomb. In 1933, at the age of twenty-five, Sobolev became the youngest corresponding member of the Soviet Academy of Sciences, and in 1939, the youngest full member (academician) of the Academy. After joining the Bolshevik Party in 1940, Sobolev was appointed as the deputy director of the Institute of Atomic Energy in 1943 and contributed to the construction of the first Soviet atomic and hydrogen bombs. With this in mind, Lyapunov and Kitov arranged to visit Sobolev at his dacha in Zvenigorod, an hour west of Moscow, where, after discussing the draft, Sobolev offered his name as coauthor. Although it is not known how much he contributed to the article, Sobolev repeatedly and publicly defended cybernetics in the late 1950s.[67]

Sometime in 1952, Kitov and Lyapunov visited the editorial staff of *Problems of Philosophy*. For unknown reasons, the editors agreed to publish the article, asking only that they receive permission from the Communist Party first. We may speculate on why the editors agreed to publish on a forbidden topic. *Voprosi Filosofii* continued to publish anticybernetic material for several years, so one might suppose that the editors thought permission would not be granted, thus shifting the blame for the rejection onto higher authorities. It is equally possible that the editors agreed to publish the article out of genuine enthusiasm to encourage intellectual debate during Khrushchev's thaw. Regardless, the editors sent Lyapunov and Kitov to meet with representatives in the science division of Staraya Square, an administrative wing for the Communist Party in downtown Moscow. The administrators heard their case, asked some questions, and then concluded: "We understand: it is necessary to change the relationship to cybernetics, but an instantaneous split is not possible: before the article can be published, it would make sense to do several public reports."[68] Lyapunov and Kitov spent 1953 and 1954 carrying out tacitly approved public lectures and private workshops, and Lyapunov began hosting in his home a circle of colleagues to discuss cybernetics that lasted over a decade.[69]

At once an introduction, a reclamation, and a creative translation of Wiener's *Cybernetics*, Kitov, Lyapunov, and Sobolev's feature article, "The Main Features of Cybernetics," danced a deliberate two-step. First, it attempted to upgrade cybernetics to parity with other natural sciences by basing an ambitiously comprehensive theory of control and communication almost exclusively on Wiener's 1948 book (although these early Soviet cybernetics made notably less of the field as an applied science and more of it as a universalizing theory than did Wiener). Second, it retooled Wiener's conceptual vocabulary into a Soviet language of science. Gerovitch details the translation of their terms: "What Wiener called 'the feedback mechanism' they called 'the theory of feedback' ... 'basic principles of digital computing' became 'the theory of automatic high-speed electronic calculating machines'; 'cybernetic models of human thinking' became the 'theory of self-organizing logical processes.'"[70] In fact, the coauthors used the word *theory* six times in their definition of cybernetics to emphasize the theoretical nature of the new science, possibly as a way to avoid having to discuss the political implications of introducing a practical field of human-machine applications into a society well suited to adopt them.

The coauthors also integrated and expanded the stochastic analysis of Claude Shannon's information theory while simultaneously stripping Wiener's organism-machine analogy of its political potency.[71] Wiener's core analogies between animal and machine, machine and mind were stressed as analogies—or how "self-organizing logical processes [appeared] *similar* to the processes of human thought" but were not synonyms. At the same time, the article scripts his language of control, feedback, and automated systems in the machine and organism into the common language of *information*, or Shannon's mathematical theory of communication. For Kitov, this "doctrine of information" took on wholesale the task of universalizing statistical control in machines and minds. It did so by preferring the "automatic high-speed electronic calculating machine" (that is, computer) to Wiener's original base analogy for cybernetic comparisons—the servomechanism. The servomechanism is an automatic engineering device used in a larger mechanism to correct, using error-sensing negative feedback, that mechanism's performance: examples could include the steam engine governor, modern cruise control in cars, or, in Wiener's case, antiaircraft fire control mechanisms controlling a gun and its gunner.[72] Despite the coauthors' efforts to silence the social implications of the theory, computer algorithms added a further layer of technical complication to Wiener's feedback mechanisms, even as their neuronal analog to electronic switches quietly implied opening new research horizons in human-computer interaction, robotic prosthetics, and cyborgs. By formulating the science in terms of cutting-edge computers, not servomechanisms, the coauthors propelled the Soviet cyberneticist and his computer into the front lines of the escalating space and technology race. Thus, conceiving of the computer as a general regulating machine for any control systems, the Soviet formulation of cybernetics focused on computational systems from the start—a generalized step away from Wiener's interests in communication and control in concrete entities of "the animal and the machine."[73] Although computers were not common in the Soviet Union until decades later, to this day, the Russian word for *cybernetics, kibernetika*—together with its heir *informatics*, or *informatika*—remains a near synonym for the English field of computer science.

Computers at the time were new media in the sense that few people agreed how to talk about them: the computer in Russian in the 1950s and 1960s went by the bulky description "automatic high-speed electronic calculating machine."[74] Frequent use of the term mercifully introduced the abbreviation *EVM* (*electronnaya vyichislitel'naya mashchina*, or "electronic calculating machine"), which stuck through the 1960s and 1970s. Only under Gorbachev's *perestroika* in the 1980s did the now nearly ubiquitous English calque *komp'yuter* replace the term *EVM*.[75] The unwieldiness of the original Soviet term underscores the perennially renewable nature of the discursive contest that makes computers more or less new. Because the coauthors were sensitive to how language, especially foreign terms, packs in questions of international competition, the coauthors attempted to keep their language as technical and abstract as possible, reminding the reader that the cybernetic mind-machine analogy was central to the emerging science but should be understood only "from a functional point of view," not a philosophical one.[76]

The technical and abstract mathematical language of Wiener's cybernetics thus served as a political defense against Soviet philosopher-critics and as ballast for generalizing the coauthors' ambitions for scientists in other fields. They employed a full toolbox of cybernetic terminology, including signal words such as *homeostasis, feedback, entropy, reflex,* and *the binary digit*. They also repeated Wiener and Shannon's emphases on probabilistic, stochastic processes as the preferred mathematical medium for scripting behavioral patterns onto abstract logical systems, including a whole section that elaborated on the mind-machine analogy with special emphasis on the central processor as capable of memory, responsiveness, and learning.[77] Wiener's call for cyberneticists with "Leibnizian catholicity" of scientific interests was tempered into its negative form—a warning against disciplinary isolationism.[78]

On the last page of the article, the coauthors smoothed over the adoption of Wiener, an American, as foreign founder of Soviet cybernetics by summarizing and stylizing Wiener's "sharp critique of capitalist society," his pseudo-Marxist prediction of a "new industrial revolution" that would arise out of the "chaotic conditions of the capitalist market," and his widely publicized postwar fear of "the replacement of common workers with mechanical robots."[79] A word play in Russian animates this last phrase: the Russian word for *worker*, or *rabotnik*, differs only by a vowel transformation from *robot*, the nearly universal term coined in 1927 by the playwright Karel Capek from the Czech word for "forced labor."[80] The first industrial revolution replaced the hand with the machine, or the *rabotnik* with the *robot*, and Wiener's science, the coauthors dreamed, would help usher in a "second industrial revolution" in which the labor of the human mind could be carried out by intelligent machines, thus freeing, as Marx had intimated a century earlier, the mind to higher pursuits. "Automation in the socialist society," the coauthors wrote in anticipation of Khrushchev's declaration at the 1956 Congress, "will help facilitate and increase the productivity of human labor."[81] Although Stalin had found no use for Wiener's sounding of a "new industrial revolution," these mathematicians had found and refashioned in Wiener an American critic of capitalism, a founder of a science that was fit to sound the Soviet call for the "increased productivity of labor."[82]

Given this explicit adoption of Wiener into the Soviet scientific canon, it is surprising to note that the coauthors quoted only one line from any of his works. That line reads: "Information is information, not matter and not energy. Any materialism that cannot allow for this cannot exist in the present."[83] By distinguishing between information, energy, and matter, Wiener skips across two recent paradigm shifts in modern physics—first, from a Newtonian physics of matter to an era of thermodynamics and Bergson and second, from the thermodynamics of energy to a new but related paradigm of information science and Wiener's cybernetics. For many in the West, this quote meant that information is nothing but information, a value-neutral statistical measurement on which to rest objective science and the search for computable truth. The technical meaning was the same for their Soviet counterparts, but it also meant something more. By singling out Wiener's alliance of materialism and cybernetics, the coauthors implied that Wiener had in mind a position that was amendable to the official philosophy of Soviet science—the dialectical materialism of Marxism-Leninism. If dialectical materialism did not update itself for the information age, it could not exist. The same quote also leaves open the opportunity that the coauthors were lobbying for—that Soviet dialectical materialism *could* allow for information to be information in its fullest cybernetic or stochastic sense. The quote thus renders Wiener as a sort of foreign prophet announcing a dialectical materialist science of information science, a science whose present materialism could only be fully Soviet. With these ritual words, the coauthors wed cybernetics to Soviet ideology and dialectical materialism to the cybernetic information sciences. The success of this "important new field" of Marxist-Leninist information science, they contended, hung on the call to action that was voiced by its American originator.

The coauthors also buttressed Wiener's ideas of neural processing with reference to the great Soviet scientist Ivan Pavlov, whose original theory of conditioned reflexes in human psychology was derived from a telephone electrical switchboard, a communication machine with ideal cybernetic resonance.[84] Finally, the coauthors concluded the article in a ritual flourish of Orwellian newspeak that was common to academic writing at the time, calling for a battle against the capitalists who "strive to humiliate the activity of the working masses that fight against capitalist exploitation. We must decisively unmask this hostile ideology."[85] After years of anti-American, anticybernetic positions, they were the first to voice an anti-American, procybernetic position in the Soviet press. In the mid-1950s, the tone of subsequent arguments began to distinguish between the capitalist use of cybernetics, which was flatly condemned, and cybernetics in general, thus creating space for the argument that the socialist use of cybernetics might not only be possible but even preferable.

"The Dark Angel": Ernest Kolman's "What Is Cybernetics?"

Whatever rhetorical flourishes Kitov, Lyapunov, and Sobolev mustered, the strongest ideological support for their newfound procybernetic position lay in the article that immediately followed their publication in the same journal, Ernest Kolman's piece "What Is Cybernetics?" ("Chto takoe kibernetika?"). A loyal Bolshevik, an active ideologue-philosopher, and a failed mathematician with a long and bloody personal history of attacking nonorthodox mathematicians, Kolman makes a somewhat surprising candidate for the first ideological defender of Soviet cybernetics.[86] Among other ideological offenses that he appears to have committed, he seems to have done the most harm to the founders of the Moscow School of Mathematics, a powerful school in imperial Russia and the Soviet Union. He excoriated them for their nonatheistic commitment to a fascinating intellectual alliance between French set theory and a Russian Orthodox name-worshipping mysticism. (Their scandalously religious observation began by noting that both infinity and God could be named but not counted.)[87] Kolman was once dubbed "one of the most savage Stalinists on the front of science and technology" for his tireless defense of Lysenko's biology (which is now remembered as the Soviet pseudo-scientific alternative to classical genetics).[88] Some Soviet commentators feel that Kolman's diatribes kept the mathematician Andrei Kolmogorov in the 1940s from beating Wiener—the two are often compared as intellectual peers—to formalizing the link between biology and mathematics. Kolman was sensitive to political attacks and had a genuine interest in the history of science and a knowledge of four or five languages. A formidable opponent, he was sometimes known among his detractors and victims as the "dark angel."[89]

Despite such a body count, Kolman's role as self-elected guardian of cybernetics was not the first time he had deviated from an ideologically orthodox line of philosophy. He had spent time in a Stalinist labor camp after World War II for straying from the party line in his interpretation of Marxism. Just before he died in 1982, he published the book *We Should Not Have Lived That Way*, in which he reflected on his own past transgressions: "In my time I evaluated many things, including the most important facts, extremely incorrectly. Sincerely deluded, I was nourished by illusions which later deceived me, but at that time I struggled for their realization, sacrificing everyone."[90] This context makes Kolman's defense of cybernetics more surprising: why would an embittered former mathematician with a track record of decimating pseudo-scientific mathematical theories come to the defense of cybernetics in 1953? Was his role as the first ideologue to defend Soviet cybernetics an act of penitence or another cardinal sin?

Kolman began his eleven-page promotional history by outlining over a century of international cybernetics, beginning with the French mathematician, physicist, and philosopher André-Marie Ampère in 1843 and moving to "Russian and Soviet scientists, [such as] Chernishwev, Shorin, Andropov, Kulebakin, and others."[91] Kolman called Wiener "one of the most visible American mathematicians and professor of mathematics at Columbia University" and the one who "definitively" formalized cybernetics "as a scientific sphere," in a veritable shout of praise for the time.[92] In fact, Wiener had been appointed at MIT, not Columbia, since 1919, but Kolman may have introduced the mistake on purpose: Columbia University stood out to Soviet observers among American universities at the time for its Russian studies center, the Harriman Institute, which had been a favorite target of McCarthy, so by connecting Wiener to Columbia, not MIT, perhaps he softened his image in the eyes of Kolman's peer philosopher-critics.[93] In any case, Wiener occupies the sixth through the ninth paragraphs of Kolman's ideological support piece, which signals a second witness of Wiener's adoption into the vanguard of Soviet cybernetic historiography.

Having set up Wiener as the foreign founder of Soviet cybernetics in the article, Kolman promptly invented a Soviet prehistory to the science that broadened and colored the ambition of cybernetics to match Marxism-Leninism. Sensitive to the many eastern European origins of cybernetic-style thinking, Kolman's narrative assimilates cybernetics into a longer history of computational machines, including Ramon Llull in 1235, Pascal in the mid-1600s, the engineer Wilgott "Odhner of St. Petersburg" (and not Stockholm, Wilgott's native city), and the late nineteenth-century mathematicians A. N. Krilov and P. L. Chebishev. He then discussed how the Soviet mathematicians Andrei Markov Jr. (a constructivist mathematician who later became a leading cyberneticist), N. C. Novikov, N. A. Shanin, and others advanced the last hundred years' worth of precybernetic work in Russian.[94] Kolman's internationalism allowed two people west of Berlin to slip into his history—Norbert Wiener and Nikolai Rashevsky, the first Pavlov-inspired biomathematician and a Russian émigré at the University of Chicago.

Thus, the battle to legitimize Soviet cybernetics began internally and was fought against by and among Soviet philosopher-critics, the vanguard and police of ideological debate in Soviet discourse. Both procybernetic articles (especially Kolman's) were loaded with discursive tactics that were meant to protect cybernetics from counterattacks, so much so that, even in pronouncing it, the first act of Soviet cybernetics partook in cold war game-theoretic strategies. In the first public defense of cybernetics, which was a lecture given at Moscow State University in 1954, Kolman notes that "it is, of course, very easy and simple to defame cybernetics as mystifying and unscientific. In my opinion, however, it would be a mistake to assume that our enemies are busy with nonsensical things, that they waste enormous means, create institutes, arrange national conferences and international congresses, publish magazines—and all this only for the purpose of discrediting the teachings of Pavlov and dragging idealism and metaphysics into psychology and sociology." By imagining enemies as rational actors, not pseudoscientific bourgeois, a cybernetic worldview provides its own first defense: "There are more effective and less expensive means than the occupation with cybernetics," Kolman the philosopher-critic continues, "if one intends to pursue idealistic and military propaganda.[95]

Kolman employed the logic of reversing the rational enemy that was implicit in all Soviet cybernetic strategy to save the fledgling movement from future Soviet critics. Kolman invites his Soviet listeners to consider cybernetics from the perspective of an economically rational American scientist.[96] We should imitate the enemy, Kolman reasoned, because we can infer that the enemy knows something we do not, for he is occupied with something we do not understand. To its participants, cybernetics took initial shape in a militarized discourse of the postwar and cold war.

Like Kolman, the coauthors Sobolev, Lyapunov, and Kitov also pre-empted the reactions of the Soviet philosophers, rebuffing them for "mis-interpreting cybernetics, suppressing cybernetic works, and ignoring the practical achievements in this field."[97] The coauthors flipped the reaction-ary argument that was sure to follow (that Soviet cybernetic defenders were "'kowtowing' before the West") by insisting that "some of our philosophers have made a serious mistake: without understanding the issue, they began by denying the validity of a new scientific trend largely because of the sen-sational noise made about it abroad."[98] In a concluding flourish, the coau-thors conspired:

One cannot exclude the possibility that the hardened reactionary and idealistic in-terpretation of cybernetics in the popular reactionary literature was especially orga-nized to disorient Soviet scientists and engineers in order to slow down the develop-ment of this new important scientific trend in our country.[99]

Thus, the coauthors held, the critics of cybernetics, not its proponents, should be suspected of having fallen under the spell of the cold war enemy. To recognize the contributions of the enemy without opening themselves to attack, they heaped suspicion on suspicion, insinuating that instiga-tors abroad had somehow organized the ideological critique of cybernetics within the Soviet Union. Although it is unlikely that the coauthors genu-inely believed that their discovery of cybernetics came in spite of the efforts of American spies and agents, this kind of argument nonetheless won inter-nal wars of words.

Soviet cyberneticists were not alone in employing this strained logic. If Wiener was right in arguing that information arms all its possessors equally, double heaps of suspicion may support an ultrarational strategy that strains toward the irrationality found across cold war discourse. Kolman's counter-defense of cybernetics against other Soviet critics, for example, resembles a game-theoretic scenario in which (like the policy of mutually assured destruction) both parties seek to settle their disagreements in order to avoid a larger collective loss.[100] The basic logic of this cybernetic worldview, asserts historian Peter Galison, is to adopt the logic of the "enemy Other" and to preempt and predict the behavior of the intelligent and rational foe to the point where the positions are reversed and foe and friend become indistinguishable.[101]

Cybernetics—like its sister disciplines of game theory, information theory, and others—appears as a method for rationalizing the enemy, dis-tributing structural strategy evenly across opponents and flattening the chances that an enemy will have to take strategic or logical advantage over an ally. Perhaps nowhere is this as clear as in the Soviet defense of cybernetics itself, except that in Kolman's case, the enemy to defend cybernetics against was his own kind. At first rejected for its American sources, Soviet cybernetics took shape not as a Soviet reaction against the American enemy but as a circular defense of Soviet mathematicians against their own philosopher-critics.

A "Complete Cybernetics": Toward a Totalizing Plurality

The efforts of Sobolev, Lyapunov, Kitov, and Kolman in print and in public lecture, combined with the intellectual weight of preeminent mathematician Andrey Kolmogorov and high-ranking administrator and engineer Aksel' Berg, led to the establishment of the statewide Council for Cybernetics in 1959, which in turn promised cybernetics a base for significant growth as an institutional field in the early 1960s. By 1965, however, it was still not clear in which direction this new science would lead. Would it distribute the powers of the Soviet state among its participants more equitably and flexibly? Or would it consolidate power still further? In 1965, an American visitor feared the worst: after visiting a facility with an evident generation gap between "all the young, recent graduates of technical higher schools" who were interested in computers and "the older bureaucrats," he prophesied that "a turnover in generations in the Soviet administration" could lead to a "computer revolution" that "may enormously increase the effectiveness of formal communication channels." The "modernization of communication may have the paradoxical effects," the American observer fretted, "of actually enhance[ing] totalitarian control by making a fully centralized network of administrative communications channels really feasible."[102]

Between 1960 and 1961, the popular press began heralding computers as "machines of communism" and engineer admiral Aksel' Berg, then director of the Council of Cybernetics, launched the first of a series of volumes entitled *Cybernetics: In the Service of Communism*.[103] This series stirred emotions among Western observers. One American reviewer noted with concern in 1963, "If any country were to achieve a completely integrated and controlled economy in which 'cybernetic' principles were applied to achieve various goals, the Soviet Union would be ahead of the United States in reaching such a state." The reviewer also picked up on the burgeoning interest in economic cybernetics, stating that "a significantly more efficient and productive Soviet economy would pose a major threat to the economic and political objectives of the Western World.... Cybernetics, in the broad meaning given it in the Soviet Union," he concluded with a flare, "may be one of the weapons Khrushchev had in mind when he threatened to 'bury' the West."[104]

The Central Committee began publicly promoting cybernetics along similar lines in 1961 at the Twenty-second Party Congress as "one of the major tools of the creation of a communist society."[105] First Secretary Nikita Khrushchev himself promoted a far-reaching application of cybernetics: "it is imperative," he declared to the Congress, "to organize wider application of cybernetics, electronic computing, and control installations in production, research work, drafting and designing, planning, accounting, statistics, and management."[106] Central Intelligence Agency (CIA) sources noted similar enthusiasm at an All-Union Conference on the Philosophical Problems of Cybernetics held in June 1962 in Moscow, which included "approximately 1000 specialists, mathematicians, philosophers, physicists, economists, psychologists, biologists, engineers, linguistics, physicians."[107] The conference adopted an official, if vague, definition of *cybernetics* as "the science which deals with the purposeful control of complex dynamic systems."[108] The most ambitious of these complex dynamic systems, the Party leadership's support seemed to imply, would be the Soviet Union itself.

The looming menace of a well-organized, cybernetic self-governing socialist enemy worried some American observers as well. During the John F. Kennedy administration, members of the intelligence community agitated against the perceived looming peril of Soviet cybernetics. John J. Ford, then a Russian specialist in the CIA and a future president of the American Society for Cybernetics, was responsible for several alarm-generating reports on Soviet cybernetics, which had already grabbed Attorney General Robert F. Kennedy's attention. One fateful evening in the fall of 1962, Ford gathered with President John F. Kennedy's top men to discuss the impending peril of Soviet cybernetics, only to have his meeting interrupted by the announcement that surveillance satellites had just uncovered photos of Soviet missiles in Cuba.[109] By the time the dust settled after the Cuban missile crisis, Soviet cybernetics no longer agitated the administration, which had reviewed the science and did not deem it an urgent threat. It is a strange twist of history, then, that the international crisis that is considered the zenith of cold war hostility (the Cuban missile crisis) also defused and derailed mounting American anxieties about the "Soviet cybernetic menace."[110]

Although U.S. and Soviet intelligence officers alternately fretted about or enthused over the possibilities of a cybernetically coordinated Soviet power, the facts about the practical debates among Soviet scientists point in a very different direction. Soviet cybernetics, for all its talk about self-governance, was anything but. Berg's series *Cybernetics: In the Service of Communism* produced heated debate and fierce divisions among prominent mathematicians in the Soviet Union.[111] In contrast to the CIA's fear of a mounting, unified platform of Soviet cybernetics, cybernetic talk swelled the internal discord among mathematical cyberneticists, painting a picture instead of an intellectually fractured front. Leading Soviet cyberneticists defined the field in dramatically different terms: Kolmogorov fought to claim information as the base of cybernetics, Markov preferred probabilistic causal networks, Lyapunov set theory, and Iablonsky algebraic logic. In 1958, only three years after their initial article, Kitov, Lyapunov, and Sobolev published an article outlining four new definitions of cybernetics in the Soviet Union, emphasizing the dominant study of "control systems," Wiener's interest in "governance and control in machines, living organisms, and human society," Kolmogorov's "processes of transmission, processing, and storing information," and Lyapunov's methods for manipulating the "structure of algorithms."[112] According to researchers, loose groups of cybernetic thought consolidated around leading cyberneticists such as Lebedev, Berg, Lyapunov, Glushkov, Ershov, and others.[113]

Although some scientists contended that the virtue of cybernetics lay in its capacious tent of competing foundations, not everyone felt that the new field should contain multitudes. Igor Poletaev, a leading Soviet information theorist and author of the 1958 book *Signal,* an early work on Soviet cybernetics, argued in 1964 against any plastic understanding of cybernetics. He legitimated his call for disciplinary coherence by invoking its foreign founder, Norbert Wiener, claiming that "'terminological inaccuracy' is unacceptable, for it leads and (has already led) to a departure from Wiener's original vision of cybernetics toward an inappropriate and irrational expansion of its subject."[114] "As a result," Poletaev continued, "the specificity of the cybernetic subject matter completely disappears, and cybernetics turns into an 'all-encompassing science of sciences,' which is against its true nature."[115] The geneticist Nikolai Timofeef-Ressovsky, whose life and work was praised and persecuted under the regimes of both Hitler and Stalin, once put the same sentiment in lighter terms. In correspondence with Lyapunov, he replaced the Russian word for *confusion* or *mess* with the term *cybernetics*, joking about his having once placed a letter in the wrong envelope as a "complete cybernetics."[116] In Timofeef-Ressovsky's witticism, we uncover a fitting rejoinder to those enthused and worried that a complete cybernetics might mean a unified Soviet information science and society.

To put it both precisely and audaciously, the term *cybernetics* should be used in the plural, and perhaps the only stable sense of cybernetics is the adjectival form, *cybernetic*, an adjunct to anything that its users see fit to apply it. From the point of view of the central committee that organized cybernetics institutionally, Soviet cybernetics, at the peak of its reach, appears both comprehensive and pluralist. It was a complete mess, as Timofeef-Ressovsky jested. In the late 1960s, the Academy of Sciences of the USSR promoted cybernetics into an entire division, one of four divisions comprising all Soviet science.[117] The remaining three (noncybernetic) divisions—"the physico-technical and mathematical sciences, chemico-technical and biological sciences, and social sciences"—could without much conceptual violence be read as subfields of the Siberian-sized Soviet cybernetic science.

The Soviets were not alone in the instinct to universalize science, although the ideological organs of the state excelled at promoting such discourse. The ecumenical commitment and a totalizing mission to stitch together the mechanical, the organic, and the social often were attributed to their foreign founder. In 1948, Wiener attempted to analogize (in the subtitle to his 1948 *Cybernetics*) "the animal and the machine" and concluded with a comment about the insufficiency of cybernetic methods for social sciences. Nonetheless, two years later, in 1950, Wiener published a popular version called *The Human Use of Human Beings*, whose subtitle belies his earlier caution: "cybernetics and society."[118] Still, the instinct to institutionalize his intellectual catholicity was clearly native to the Academy of Science, which originally categorized cybernetics into eight sections, including mathematics, engineering, economics, mathematical machines, biology, linguistics, reliability theory, and a "special" military section.[119] With Aksel' Berg's sway over the Council on Cybernetics, the number of recognized subfields then grew to envelop "geological cybernetics," "agricultural cybernetics," "geographical cybernetics," "theoretical cybernetics" (mathematics), "bio-cybernetics" (sometimes "bionics" or biological sciences), and, the most prominent of the Soviet cybernetic social sciences, "economic cybernetics" (discussed in later chapters).[120]

By 1967, the range of cybernetic sections enveloped information theory, information systems, bionics, chemistry, psychology, energy systems, transportation, and justice, with semiotics joining the linguistic section and medicine uniting with biology. Sheltering a huddling crowd of unorthodox sciences, including "non-Pavlovian physiology ('psychological cybernetics'), structural linguistics ('cybernetic linguistics'), and new approaches in experiment planning ('chemical cybernetics') and legal studies ('legal cybernetics')," cybernetics in the mid-1960s grew to an almost all-encompassing size.

Nonetheless, the runaway institutional success of cybernetics in the Soviet Union also meant that, by the time Leonid Brezhnev came to power in 1964, Soviet cybernetics could not help but slouch toward the intellectual mainstream.[121] It had to: its territory had grown so large it could not help but take up the middle of the road. The institutional growth of cybernetics outran the intellectual legs supporting it: the failure of cybernetics to cohere intellectually actually rested on the runaway growth of the discipline institutionally. Sloughing reformist ambitions to the side, by the 1970s, *kibernetika* signaled little more than a common interest in computer modeling that held together a loose patchwork of institutions, disciplines, fields, and topics. By the 1980s, the term *cybernetics* marked a nearly empty signifier for all the plural things to which the adjective could be attached. By the rise of Gorbachev in 1984, Soviet cybernetics had successfully accompanied and slowly integrated into a host of parallel developments. The inheritor field "informatics," the parallel revolution in military affairs, the scientific-technical revolution, and the first three generations of computer hardware (vacuum tubes, transistors, and integrated circuits) had rolled forward under the fading banner of Soviet cybernetics.[122]

Conclusion: Wiener in Moscow

This brief history of early Soviet cybernetics ends where it began, with Norbert Wiener and the foreign founding of cybernetics. In the early 1960s, travel restrictions for Americans in the Soviet Union began to slacken, and a trickle of chaperoned scientific and cultural exchanges began to flow between the two superpowers. Early among this generation of guests was Wiener, then an aging omnibus professor at MIT. In June 1960, Soviet officials warmly welcomed this American founder of cybernetics for a several-week visit to Moscow, St. Petersburg, and Kiev (figure 1.1.) After his arrival, Wiener, whose translated books were popular (albeit in edited form) in the Soviet Union, was paid royalties in cheap caviar and champagne (which apparently sat untouched in his basement) and gave invited lectures at prestigious institutes in those three cities.[123] For Wiener, it was a chance to issue a stirring warning against societies that would adopt cybernetics without the fundamental ability to correct themselves, decrying that "science must be free from the narrow restraints of political ideology." For his Soviet hosts, the visit allowed the cybernetic knowledge base to go about the regular ideological work of welcoming and canonizing a socialist saint in public memory of Soviet society and technology.[124] The effect among his colleagues in the Soviet Union and in Cambridge was electric. Reflecting on his public reception, Wiener's friend Dirk Struik, a Dutch mathematician and Marxist theoretician, captured the moment for many Soviet cyberneticists with his overstatement, "Wiener is the only man I know who conquered Russia, and single-handed at that."[125] We may claim that by this process Wiener became known as a foreign founder of Soviet cybernetics. In *Democracy and the Foreigner*, political theorist Bonnie Honig introduces the idea that an iconic "foreign founder," or an alien recruited for a project that he or she unsettled, often plays a role in the many political narratives of identity formation: the kingdom of Oz has its Dorothy of Kansas; the House of David has a Moabite grandmother, Ruth; the American colonies were united by the belief that they were no longer British; Europe now traces its origins to ancient Greece, which was first a Roman idea. Eastern Europe abounds in similar stories: Russia originates in ancient Rus', now in Ukraine; the Ukrainian national anthem claims brotherhood with the Cossack; and the Polish national anthem praises Lithuania.[126] That Soviet cybernetics identified Wiener as foreign founder is in context nothing new. After all, no native can found his or her identity. There is no identity

Figure 1.1
Norbert Wiener with Aleksei A. Lyapunov in Moscow, 1960.
Courtesy of Boris Malinovsky.

without a founder, and because founders precede identities, all foundations must be laid by what must appear post fact as foreigners.

Wiener's renown in the former Soviet territories has outlasted his memory in the English-speaking world. When Aksel' Berg became chair of the Council on Cybernetics in 1959, he made sure that among the first supporting works translated were Wiener's. Over fifty years later, nearly all of Wiener's major works have since been translated into Russian and retain their relative popularity, long after his legacy has faded in the English-speaking world, except recently among historians of science.[127] Wiener's 1948 *Cybernetics, or Control and Communication in the Animal and the Machine* was translated into Russian in 1958 and reissued in 1968 and again in 1983, one more printing than in English. In 1958, his *Human Use of Human Beings: Cybernetics and Society* was abridged and translated as *Kibernetika i obscheshtvo* (*Cybernetics and Society*). Based on the lectures he gave while visiting Moscow, he published a 1962 article "Science and Society" in the preeminent journal *Problemy Philosophii* (*Problems of Philosophy*). His autobiographies *Ex-Prodigy* (1953) and *I Am a Mathematician* (1956) were translated in 1967. And his final collection of lectures, *God and Golem, Inc.: A Comment on Certain Points in Which Cybernetics Impinges on Religion* (1964), was translated as *Tvorets i robot* (*Creator and Robot*) in 1966 and reissued in 2003.

As a testament to the staying power of Wiener as an iconic foreign founder figure, Wiener's semiautobiographical novel *The Tempter* was translated in 1972, eight years after his death. His short piece of fiction, "The Brain," which is hard to find in English, was translated in 1988. And his 1951 article "Homeostasis in the Individual and Society" appeared in Russian in 1992, just after the turbulent collapse of Soviet society. Bookstores in Moscow continue to offer new editions of Wiener's works to this day. His oeuvre has also migrated online unevenly: all aforementioned works in Russian are freely available for download online, compared to only one work in English, *God and Golem, Inc.* Given all this, it may not be a stretch to assert that, with the visit of an American founder of cybernetics, the son of Leo Wiener, an émigré from Byelostock and founder of Slavic studies in America, Norbert Wiener was christened no less than a Soviet prophet returning home.[128]

Yet if Wiener were a prophet, he would be the kind whose stinging calls to repentance went ignored both at home and abroad. He pressed for removing ideology from science just as the political winds, in the early 1960s, were shifting toward ideological reconsolidation and recentralization under Brezhnev. The case of Wiener in Moscow is interesting, then, not merely for biographical or historiographical reasons but also as a synecdoche for the larger Soviet experience with cybernetics. The cybernetic technological apparatus brought with it a promise of systemwide structural reform, and although that reform was never fully realized, the technological apparatus was. Cybernetics accompanied the transformation of Soviet society into an already networked information society, although it did so without bringing about the intended social, organization, and technological reforms and self-governance. The early nationwide cybernetworks explored in subsequent chapters are central to understanding the Soviet experience and the unintended political consequences of sociotechnical and technocratic reforms.

A glance at the history of early Soviet cybernetics might at first steer readers to think that technocratic sciences are politically neutral, capable of adapting to whatever the political discourse of the day is, whether Stalin's rejection, Khrushchev's reform, or Brezhnev's reconsolidation of technocratic science. Yet this is not the case: claiming technocratic neutrality itself is a consequential political posture that often is filled by whatever the politics of status quo at the time and place are. The nationwide networks created to save the flagging economy and technical data infrastructures discussed in later chapters are presented as socially neutral technocratic solutions to social problems—and yet that position of neutrality proved to be a veiled form of ideational investment. Considered generally, the cybernetic goal of controlling and regulating information systems in abstract and supposedly neutral mathematical terms appealed to post-Stalinist scientists who were fed up with political oppression. Cybernetics struck Moscow-based bureaucrats and party officials as a politically feasible way forward in preserving the centralized state as an information system without the abuses of Stalinism.[129] Behold the promise of control without violence and of a socialist information society liberated from its stained past by the neutralizing politics of computation.

Others promised technological improvements without politics long before the onset of computers and digital media. Soviet discourse of what James Carey called the "electric sublime" begins with Lenin's famous 1920 statement that "Communism is Soviet power plus the electrification of the whole country," perhaps the highpoint of the Soviet reputation in the West as well as a memorable declaration of the Soviet Union's commitment to achieve social progress through technological modernization.[130] Soviet cybernetic discourse built actively on that tradition—particularly that of the Soviet digital economic network projects, which, like Lenin's electrification (or GOERLO) project, promised to rework the technological infrastructure of the whole country—the factories, the grids that united them, and the giant hydroelectric and computer stations that powered them. The cybernetwork projects integrated and updated a longer tradition of the industrialist, Taylorist megaprojects that marked the Soviet electrical age.

The cybernetic lexicon also resonates richly with native Soviet discourse. Before Wiener cemented that hardy word as central to cybernetic systems, *feedback* occupied a prominent position in the Soviet political imagination of itself as a "socialist democracy," a kind of complex social entity sustained by Pavlovian mechanisms of stimulus and response and control and cooperation between rulers and masses.[131] With little work, the term *noise reduction* came to stand for a technical synonym for continuing political censorship in the Soviet Union. Moreover, Wiener's twinning of the modern laborer with an automaton echoed of Stalin's attempts to make Soviet labor and industry efficient with the scientific management techniques of Taylorism. Wiener's theories of systematic information control and communication, once translated into Russian, appeared to be a recuperation of ideas that already were well understood.[132]

Perhaps this history of Soviet cybernetics is most helpful not for what it says about cybernetics but for what the discursive pliancy of cybernetics allows us to see in Soviet society. As a term, *cybernetics* served as a flexible semantic placeholder for a more widely held article of faith about the promise of technocratic governance aided by computer in post-Stalinist science and society. As a history, the several-step process of the Soviet rejection, rehabilitation, adoption, and adaptation of a new foreign discipline reveals less about cybernetics than it recapitulates the preexisting political dynamics of Soviet discourse—the debate patterns, rituals of discourse, strategies for intellectual defense, alliance forging, institution building, the political whims of Moscow, and other everyday dynamics. Backlit with fascinating twists, turns, and figures, the story of Soviet cybernetics presented here signals not particularly well-defined intellectual contributions but rather shows the ways that the lack of them allowed Soviet cybernetic discourse to mold to and reflect longer transformations and trends in the Soviet state's attempts to manage and control science, technology, and society.

Soviet cybernetics thus appears to be a normal science in the sense that it reveals the conflicting dynamics of underlying political, economic, and institutional practices and structures. These dynamics—the echoes of anticapitalistic public campaigns, the ritual aspects of intellectual debates and duels, the political machinations and strategies, the institutional diffusion of the computer as a specialized tool, the history of spikes of invention followed by downward-sloping plateaus of innovation and development characterizing the history of science in Russia, and the stubborn fact that the work of science takes place in prolific dialects and varied trading zones subject to the punishing pleasures of contest, prestige, and competition[133]— appeared par for the course. As the next chapter attempts to illustrate, the case of Soviet economic cybernetics challenges historians and other agent-observers of change with the suggestion that perhaps the ordinary, over-looked elements of actors, ideas, practices, and policies—including those governing everyday life in the command economy—best describe the circuitous historical course of science and social reform.

In one important sector, however, Soviet cybernetics and other information sciences were not obviously subjected to a confusion of competing motives—the Soviet military. The Red Army adopted cybernetic research methods and vocabulary, usually coded in public simply as "special research"; successfully theorized the military-technical revolution spurred by computers and associated long-range, specific-target military innovations; and maintained a competitive space and nuclear and long-range conventional warfare armaments without the internal incoherence and competition that was found in civilian sectors. So although the Soviet cybernetic-lit military technology revolution of the 1970s did not lead to application due to the political and economic incapacities of the Soviet state, the key distinction from the civilian economic sectors is that, inside the centralized military administration, real cybernetic reform was both possible and carried out in theory.[134]

In conclusion, having outlined a few sources that led to the consolidation of cybernetics in Wiener's 1948 masterwork, the Macy Conferences on Cybernetics (1946–1953), its postwar spread through France, England, Chile, and a vignette of how cybernetics became a loose techno-ideological framework for thinking through information sciences in post-Stalinist Soviet Union, I now comment on the idiosyncratic development of cybernetics across these moments in the early cold war global history. Several comparisons and contrasts draw connections to other postwar climates where cybernetics came to roost. The Soviet translation and adoption of cybernetics share with the other case studies glossed here an underlying fascination with the relationship of the mind to the machine, especially as seen in the biology and neurology of the British and Chilean cyberneticists. The mind-machine analog is a politically charged two-way street. Not only does cybernetics prompt us to think about how a logic machine (computer circuits or any other Turing machine) may function like a mind (a neural network), but it also raises McCulloch's potent possibility that subsequent neuroscience has soundly rejected: the mind (neural network) might function like a logic machine (computer circuits). This reverse comparison (that a mind is like a machine) proves particularly enduring in later discussions of the design and development of national networks. The designers of major early cold war national networks in the Soviet Union, the United States, and Chile sought, implicitly or explicitly, to model their own self-governing national networks after cybernetic neural networks. In the comparative network designs (including distributed, hierarchical, and participatory), early network scientists proposed differing images of the relationship between a network and the living body politic of the nation.

The mind analogies all share a common cybernetic impulse to analogize between information systems underlying organisms, machines, and societies. (The organizing itch of cybernetics is simply that a better-understood system can inform a less well-understood system.) Analogies are neither right nor wrong: they should be judged by their interpretive use rather than their epistemic weight. (Or as Evelyn Fox Keller once noted, the word *simulation* meant *deception* before it meant *analogical likeness*.) Given this, it is striking that each of the network architect teams at hand (Glushkov's OGAS, Beer's Cybersyn, and Baran's ARPANET) chose to analogize or model its national network project after the same basic image—the human mind, or an organic nervous system. But each of these national networks expressed the basic design analogy differently.

These early national networks projects—OGAS, Cybersyn, and ARPA-NET—were designed after different models of the (human) mind. Even though Beer and the Cybersyn project rejected previous and ongoing Soviet attempts to manage the command economy, the OGAS and Cybersyn projects pursued a national model in which the nation is likened to the body of an organism and the computer network to the nervous system that incorporates that nation's communication. The ARPANET, by contrast, inspired by McCulloch's neural network research, is analogized to a disembodied brain itself. In this case, the nation is like the brain itself: whatever organization the network serves constitutes its own neural network. To oversimplify, Baran foresaw a national state network simulating a brain *without* a body, while Glushkov (and Beer) anticipated a network nation simulating a body *with* a brain—a government in touch with its people.

As cognitive philosophers have submitted, analogies of (a nation as) an embodied mind and a disembodied brain work very differently. Although Soviet scientists were understandably wary of overbold political proclamations, the OGAS design reaffirmed the self-conception of the Soviet state as a decentralized hierarchical heart of the Soviet nation. In a colossal nation, workers were to be incorporated and animated by planning that emanated from the central processing unit, or social brain, in Moscow. That state would not be simply centralized and top-down. In the OGAS design, the network would serve as a nationwide nervous system that responded to and adjusted in real time to local events and maintained dynamic balance through complex feedback loops with its internal and international information environment. This metaphor was both materialist and idealist—materialist in that it grounded the nation in the industrial and economic realities already on the ground and idealist in that it ignored the fact that the economy did not behave like a healthy or single body (but instead like an environment for nonsymbiotic competition over bureaucratic positioning).

Also consider how the U.S. ARPANET analogy of the nation as a disembodied brain, although articulated here for the first time to my knowledge, has already been inscribed many times. Most often the interpretations smack of triumphalist political overtones. Seeing the nation (network) as a brain, not a body, signifies that the United States is conceived as an organ for knowledge work, not physical labor; that its civilian communication networks imagine its citizens, not the state, as the democratic decision-making mechanism for the nation; that those citizens exist in peer-to-peer relationships where, like nodes in a distributed network, each may act and interact with her neighbor as equals; and that (particularly common in digital libertarian discourse) the computer network itself constitutes the higher order of technological freedom that is necessary for the natural emergence of a more robust political order. (When Baran described distributed networking, his word was not *robust* but *survivable* because his network was to survive nuclear attack, which puts a less optimistic spin on things.) Baran, we might assert, was acting in the libertarian tradition by espousing the organic nation as a marketplace of individuals dating back to Herbert Spencer.[136] Or perhaps Baran designed the ARPANET after the image of the state as an enlightened social brain, channeling the American progressive notion of the state (or any other depository of organized intelligence, including the news-reading public, schools, universities, and scientific laboratories) as a "social sensorium" dating back to John Dewey and Charles Horton Cooley.[137]

With enough imagination, the analogy of the national network to a human mind can serve almost any end, such as the engine of a sensing being interacting in a mediated environment, a nervous system animating a living body, or the gray matter filling a skull. Perhaps the reason that these cyberneticists populated their analogies with the human mind was (to paraphrase a leading neuroscientist) simply the fact that humans like to believe that the human mind is the most complicated thing in the universe, even though this idea is probably no more than the brain's opinion about itself.[138] The point is that this analog, like all others, is contentless. It has no right or wrong, and the work that it does for us is work that we do to ourselves. The stories we tell ourselves about our networks reveal more about us—the spinners of modern-day network rhetoric—than it does about the network itself.

Finally, this chapter summarizes how, in its early adoption period, early Soviet cybernetics muted but did not erase politically potent mind-nation-network questions with language that was deliberately more technocratic and theoretical perhaps than that of cyberneticists in other countries. Although no surprise, talk about cybernetics and society took on the technical discourse of what Gerovitch calls Soviet "cyberspeak," or an ideological and discursive strategy for embedding public discussion about society in the language of technical expertise. The postwar and cold war debates about cybernetics in the Soviet Union impinged on the social implications of the new science. Perhaps the most obvious example of a technocratic approach bearing out social implications is the focus of the next chapter—the case of economic cybernetics. How, if at all, might cybernetics—or the study of communication systems that organize our bodies, machines, and societies—improve the current social, political, and economic order? How, as Stafford Beer developed in Chile, might cybernetic insights be applied to the networking of the Soviet nation in need of an economic boost? How might concerns with communication and control that were central to both the larger Soviet state and cybernetic projects play out in the crucial practice and policies of command economies? The following chapter discusses these and other questions.

2 Economic Cybernetics and Its Limits

Civilian computer networking in the Soviet Union first developed among cyberneticists who applied their science to a unique environment—the command economy. By examining the work of economic cyberneticists—a field found only in the territories of the former Soviet Union—we can begin to understand the significance of the internal economic crisis to Soviet scientists and civilians and the ways in which Soviet scientists, administrators, and policymakers in 1959 to 1963 viewed the command economy itself as a complex cybernetic organization. In this light, the same terms were used both by key Soviet network entrepreneurs to envision the first national networks as well as by the critics who condemned those projects. By reviewing the organizational theories and practices that characterize the Soviet state socialist economies, this analysis explores and begins to complicate the divide between the private markets and the public states that underlie conventional conceptions of the cold war.[1]

The command economy contained in its operations the cybernetic seeds and complex sources of its own undoing—nonlinear command and control, informal competition, vertical bargaining, and what I am calling *heterarchical networks* of administrative conflict. In this chapter, I develop these observations through a series of examples that outline the basic operations of the command economy in theory and in practice, the various schools of thought concerning economic reform (especially around the transition from Nikita Khrushchev to Leonid Brezhnev in 1963), and the political tensions that economic cybernetics tried to square itself with in an attempt to reform (often with long-distance networks) the structural contradictions underlying the practices of the command economy. These contradictions slowed efforts at technocratic economic reform and also ensured the enduring appeal of nonlinear cybernetic systems thinking.

The term *command economy* originated from the German *Befehlswirtschaft*, which was used to describe the Nazis' centralized economy and socialist economy. A command economy is one in which the coordination of economy activity is carried out not by market mechanisms but by administrative means through commands, directives, targets, quotas, regulations, and the like.[2] Karl Marx and Friedrich Engels said almost nothing about economic planning, except that it would be necessary, and Engels left the decisions to the workers.[3] They also asserted that socialism would be impossible to build in impoverished societies, which Leon Trotsky associated with tsarist Russia before fleeing to Mexico. Nikolai Bukharin foreshadowed what followed next when he said that "as soon as we make an organized social economy, all the basic 'problems' of political economy disappear: problems of value, price, profit, and the like. Here 'relations between people' are not expressed in 'relations between things,' and the social economy is regulated not by the blind forces of the market and competition, but consciously by a … *plan.*"[4] On such promises, the Russian revolution was built. Nonetheless, from 1917 until the collapse of the Soviet state in 1991, a perennial puzzle dogged the Marxist-Leninist state planners: How precisely was that plan supposed to work? How can a state command an economy?

Some tenets of the Soviet answer are clear.[5] All the means of industrial production were nationalized, and although Soviet citizens could own some "individual" (not "private") property (including houses, apartments, and automobiles), few could afford to do so.[6] The Soviet state appointed three state ministries to serve as the nation's economic brains, budget-keeper, and managers of the nation's vast property holdings and means of production—the Gosplan (State Planning Commission), the Gosbank (State Bank), and the Gossnab (State Commission for Materials and Equipment Supply). (*Gos* is short for *gosudarstvo* or Russian for *state* or *government.*) Gosbank, the central bank that prepared the state budget with the Ministry of Finance, played a transactional accounting role and the least critical role of the three.

Gosplan and Gossnab carried out crucial and different roles. Gosplan was entrusted with creating the economic plans of action—the governing documents defining the economic inputs (such as labor and raw materials), the timetable for execution, the wholesale prices, and most of the retail prices—divided into five-year increments (the so-called five-year plans). These nationwide economic plans were first rolled out from 1929 to 1933 under Stalin and ended, with one seven-year exception (1959–1965) under Khrushchev, with the twelfth plan (1986–1990), which oversaw Mikhail Gorbachev's reform policies of *uskorenie* (acceleration) and *perestroika* (rebuilding). The thirteenth five-year plan was cut short by the dissolution of the Soviet Union in 1991.

Gossnab, in contrast, was responsible for implementing Gosplan's plans by procuring and supplying producer goods to factories and enterprises and by monitoring the schedules for the production plans. Gossnab thus fulfilled the market role of allocating goods to producers and bridged the three levels of the command economy—national, regional, and local planning and production. The three-tiered model, established under Stalin in the 1930s, presents a straightforward pyramid. Gosplan sat at the top level and politically determined the national targets for each sector and industry, those targets are divided hierarchically among the midlevel of regional ministries, and they are further subdivided at the bottom level among enterprises and factories themselves.[7] If Gosplan planned it, Gossnab carried it out across all three levels—or at least that was the plan. As I lay out below, Soviet bureaucrats came to understand that at its heart, Soviet economic planning was a cybernetic process. This understanding goes a long way toward explaining the curious fact that the same state planners and economic agents later resisted attempts to implement large-scale cybercomputing networks in the Soviet Union.

The Soviet command economy grew at tremendous human, environmental, and organizational costs. In wartime, the command economy worked well enough to survive the extreme national duress of World War II, in which a devastating 26 million people or 14 percent of the Soviet population perished between 1941 and 1945. For the next few decades, Soviet gross national product grew faster than elsewhere in the world, enjoying a peak growth rate of 7 percent in the 1950s and 4 to 5 percent in the 1960s (before flattening out to a 2 percent growth rate in the 1970s and finally stalling at zero in the 1980s). In 1987, the "oppositionalist" Soviet economist G. I. Khanin estimated that Soviet economic productivity grew a total of 6.6 times (not the official claims of 84.4 times) since 1928—which by raw indices alone, is a history of economic growth similar to normal industrialized economies.[8] By far the most unforgivable and unforgettable cost to Stalin's rapid pace of economic development came in human lives. Some estimate that as many as 10 million lives were lost, many of them forced famine victims, surely among the most despairing statistics in modern history.[9]

Stalin built the state at inhuman cost, but he built it nonetheless. Under Lenin's and Stalin's leadership, the command economy modernized a preindustrial country that was run by a few into a mighty industrial power. It began in 1917 with a small group of professional socialist revolutionaries who lived in a few cities in a huge country that was 84 percent rural and whose population was over 95 percent illiterate peasants. After their

October coup, the Bolsheviks eliminated the remnants of the oppositional armies run by the tsar and the Mensheviks, among others, and developed an advanced industrialized economy that, after a couple of decades of forced modernization, helped the Allies defeat the Nazi war machine. As the cold war ensued, the Soviets, fueled in part by state paranoia and in part by scientific ambitions, maintained military parity with the United States, obtaining nuclear energy and weaponry before most of the rest of the world and pulling ahead in the space race in the late 1950s.

The political economy was also engineered to advance meaningful civilian causes such as socioeconomic justice. In most empires, the revenue flows from colony to center, but in the Soviet Union the funds ran in reverse: Moscow invested more in supporting satellite republics and regions than it stripped from them. The state mandated education, raised literacy rates for millions, granted women skilled and technical positions in the workplace, and successfully exported huge amounts of natural resources, ensuring a Soviet presence on the international economic stage. In the 1920s, before the Great Depression and before the 1930s purges, the gulags, and other Stalinist abuses became widely known, most intellectuals in the West admired at least some parts of the ambitious social projects that rode the coattails of the Russian revolution.[10] Optimism glimmered again after the death of Stalin in 1953 and through the heady years of the early 1960s, when all outside indicators suggested that the magic of the command economy—a fairytale on which a repressive empire had been built—might actually be working.

Yet those backstage had a better view of the problems. The degree of information coordination between Gosplan and Gossnab—the brain for planning and the hands of the command economy—was taxing the peacetime state administration. Many things could go wrong and did. Gosplan planned it, but Gossnab did not follow through. Or Gosplan planned wrongly so that, even when properly executed, the plan did not meet the economy's needs. Rarely, if ever, did the command economy work as planned.

The problems that economic planners and practitioners faced multiplied in application. They include an accounting burden accumulated from innocent calculation errors, compounded incentives that distorted reporting, toilsome paperwork, structural inconsistencies across industry standards, prohibitively technical product orders, uncoordinated silos of the national planning apparatus already awash in pricing decisions and administrative deluge, and many other practical problems that manifested themselves to cyberneticists and other economic planners as informal competition in the command economy.

The institutional map of the command economy grew labyrinthine as the immense accounting burden—a hulking coordination problem (or in cybernetic lingo, an information-processing problem)—that was shouldered by Gosplan and Gossnab was complicated by the participation of the Ministry of Finance, the Central Statistical Administration, and the Ministry of Defense (defense is thought to have occupied as much as one quarter of the USSR's GDP in the late 1980s, although estimates vary widely).[11] In the first six months of 1962, the priority industries that produced steel tubes, mineral fertilizers, agricultural machinery, chemicals, oil, cement, and light steel fell to at least 7 percent under quota—which some critical accountants attributed to human calculation errors. A calculation error could mean too low production targets for heavy machinery one year, and too little heavy machinery that year meant cross-industry shortfalls the next.[12] Even growth, when unforeseen, spelled trouble: in 1962, it was discovered that the ongoing seven-year plan had overlooked the 1959 census data and that by July 1962, the Soviet population had grown by 4 million more than had been planned for. Khrushchev once predicted that the population discrepancy by the late 1960s would border on 15 million people unaccounted for in the young, nonproductive workforce.

Even the best-laid plans, no matter how accurately made at the ministry level, went awry from ministry to regional council to factory. On the factory floor, people encountered widespread problems when translating the quotas and orders into day-to-day operations. One factory was known for decades after the war as the producer of a series of increasingly obsolete automobile models, including the luxury government limousine known as the ZIL (the abbreviation for *Zavod imeni Likhacheva*). The ZIL factory (or Likhachev factory) received orders and quotas that were so specialized that they required especially trained experts to interpret and execute. Yet as investigators discovered in the early 1960s, only two out of sixty-four factory employees had any higher education, and twenty out of sixty-four had not completed high school. Few on the factory floor could read, yet alone fulfill, the specialized orders they received.

Every information-planning problem was also a coordination and thus organizational-institutional problem, and the further up the economic hierarchy, the more intractable the coordination problems. Even at the top of the ministries, the economic plan did not necessarily exist in a single coordinated document, and so silos of attention regimented and splintered the planning process. Consider this 1962 complaint in *Pravda* from a factory director about the determination of cost:

The department of Gosplan that drafts the production program for Sovnarkhozy [collective farms] and enterprises is totally uninterested in costs and profits.... Ask the official in the production programmed department in what factory it is cheaper to produce this or that product. He has no idea.... He is responsible only for allocating production tasks. Another department, uninterested in costs, decides the plan for gross output. A third department or subdepartment proceeding on the principle that costs must always decline and productivity increase, plans the costs, the wage fund, and the labor force on the basis of past performance. Allocations of materials and components are planned by numerous other departments. Not a single department of Gosplan is responsible for the consistency of these plans.[13]

Some ministries tried to address these problems by tailoring their own plans in-house. For example, the Ministry of Wood and Wood Processing streamlined and unified the procedural notation for its medium-sized industry. The resulting code, once formulated and printed, weighed in at a wrist-breaking eighteen hundred pages and proved incompatible with other industries.[14]

Given such perpetual misfits between plan and practice, the Soviet search for the "perfect" economic organization was, in Gertrude Schroeder's understatement, "continuous." The annals of Soviet economic planning match decade after decade of bold conceptual innovations with perpetual practical setbacks. The Gossnab ministry itself was dissolved or recreated at least once every decade after its creation in 1947. It was fully dissolved in 1953 after Stalin's death; was recreated in 1965 under Brezhnev, where it oversaw the delivery of over two thousand essential products; underwent various shufflings of responsibilities; and finally was stripped of the political supply of petroleum products in 1981.[15]

All in all, the coordination problem was simple to state yet bewildering to solve: how could the nation best manage, harmonize, and organize all the information variables, planned and otherwise, that were flowing through its economy? How, if at all, could the Soviet knowledge base— including economic cyberneticists, a group known for a taste for circular problems—hope to account for the deficiencies of accounting in the system? In 1962, the State Committee for Automation and the Institute of Statistics estimated that roughly 3 million citizens (about 1.3 percent of the 220 million total) were engaged in public accountancy, data registration, statistical and planning calculations, and other supporting information services for the planned economy and that the number was rising fast. And yet no one, outside of strong-armed national commanders under extreme

wartime conditions, could manage and execute all the operations necessary to sustain the administrative creep of bureaucrats that was necessary to
oversee the businesses, factories, and industries that were driving a national
economy. In 1962, Viktor Glushkov, the prominent cyberneticist and architect of the OGAS Project, formulated the problem that his network project
proposed a cybernetic solution for: he estimated that if the current paper-
driven methods continued unchanged, the planning bureaucracy would
grow by almost fortyfold by 1980, requiring the entire adult population of
the Soviet Union to be employed in managing its own bureaucracy.[16]

The Many Pathways and Pressures to Reform

Under Stalin's centralizing rule, the pressures to reform the cumbersome
bureaucracy of the command economy were immense yet bottled up. At
21:50 on March 5, 1953, Iosif Vissarionovich Dzhugashvili, or Stalin (a
portmantetau of Russian *stal* or "steel" and "Lenin"), died of an apparent
brain hemorrhage. His was possibly the most consequential death of the
twentieth century. It set off waves of economic reform. A mere ten days
after he died, as a salute to their deceased strong leader and out of a gut
instinct for damage control, the Politburo combined twenty-four ministries into eleven strengthened ones. The reformer Nikita Khrushchev would
have been foiled from implementing systematic administrative and economic reforms because the reforms had begun before he could ascend to
power as the new general (and then first) secretary: under his administration, a series of uneven and troubled reforms were enacted between 1956
and 1965.

By the time that Khrushchev secured power, the winds of administrative
reform were blowing in the opposite direction (even administrations follow dialectical patterns). Beginning in 1954, he began introducing dramatic
reforms to decentralize Stalin-era control over the economy, ceding some
Kremlin power to national, regional, and local subcommittees. Gossnab
itself—the national ministry for allocating goods—was dissolved from 1954
to 1964. In 1955, new laws significantly broadened the powers of regional
and local planning councils, leaving in their hands for the first time in
decades questions about their own financing, planning, capital investment,
labor and worker pay, and even some cultural and social projects. Factory
directors also took more direct responsibility in determining their factory's
planning, financing, and pay situation. In 1957, Khrushchev did away with
national industrial ministries and replaced them with regional economic
councils (called *Sovnarkhozy*). He continued to implement similar measures

over his years in power, further splintering and territorializing the single national economic administrative hierarchy into 105 economic-administrative regional councils that were overseen by ten general and fifteen union-republic ministries. The 1957 economic decentralization, Khrushchev hoped, would help streamline and localize the planning process for a monstrously complex and administratively top-heavy postwar economy with over 200,000 industrial enterprises.

The causes and effects are hard to sort out. It is estimated that of the 44.8 million workers in the Soviet Union in 1954, the administrative personnel made up 6.5 million of them, or 15 percent of the national workforce.[17] No doubt Khrushchev also harbored some hopes that his decentralizing reforms would release him from bearing sole responsibility for the health of the whole Soviet economy. And yet the reforms did not work as hoped: GNP growth plunged from 8.4 percent in 1956 to 3.8 percent in 1957, the year of Khrushchev's major reforms, and bounced around a 5 percent average until the Khrushchev-toppling disaster that was the poor harvest of 1963 (-1.1 percent decline, the only year with a negative GDP growth until the end of the Soviet Union).[18]

Cybernetic economists quickly learned a point that network theorist Alex Galloway has subsequently clarified: control does not necessarily dissipate with decentralized or distributed networks.[19] It exists in the protocols and the (network) administrators and their rulings, and planning protocols were periodically scrambled. Instead of accounting production by volume, piecemeal targets were set after decentralized planning decisions. Instead of empowering and streamlining the local economy, the decentralizing reforms enraged the old guard in Moscow against its reformer and enlarged the nation's economic administrative apparatus. The overwhelming political effects of widespread decentralization among economic administrations alienated and frustrated many party officials, exacerbating the disarray and discontent already attached to Khrushchev's volatile leadership. Nonetheless, Khrushchev's decentralization allowed for several schools of economic thought in the early 1960s to percolate into public discussion and to cohere in the debates among the top party leadership about the best path of reform.

Orthodox, Liberal, and Cybernetic Economists

In the transition from Khrushchev to Brezhnev, several camps (*schools* presumes too much order) of thought coalesced around the question of economic reform. The first camp included a generation of orthodox economists who clung to positions that many had gained under Stalin, held the then contemporary functioning of the command economy (with its pyramid bureaucracy of paper stretching across national and regional planners, accountants, and quotas), and was not only doing just fine but was the only ideologically approvable means for advancing socialist economy toward communism. The most severe of the antirevisionists had long put forward a rearguard defense of their own positions in power, which was the historical paradox that any reform to the political system would be an unacceptable deviation from the original Marxist project. Not even Marx knew how such an economy would work, and his Soviet legacy was one of continuous political economic reform. Even ultraorthodox economists had trouble persuading others that there was no room for any economic reform in the wake of Khrushchev's own economic reforms, the political thaw, and unstable economic growth. With so much at stake, no one could disagree: something had to change. The orthodox economists had to concede that there was room for debate.

The second camp took up what later was called the liberal economic position and came onto the public scene in early September 1962 with the publication of a *Pravda* article by a once obscure economics professor, Evsei G. Liberman. Liberman, the youngest son of a Ukrainian Jewish forest guard from Galicia (who eventually emigrated to New York), came to the field of economic planning relatively late at the age of thirty-seven while visiting factories in Germany in 1933. He also was responsible for introducing punched-card computers—Powers and Hollerith perforating machines—for planning in Ukrainian factories. In that 1962 *Pravda* article (titled "Plans, Profits, and Bonuses"), Liberman introduced the signature piece of his reform platform, the idea of profit reform, which he had developed in his 1956 dissertation "Profitability of Socialist Enterprise." Liberman offered up a galvanizing call for economic reform—one that would require little more than the stroke of a pencil and a slight retooling of the planning apparatus.[20] He proposed that the efficiency of an economic enterprise should be measured by its profitability rather than its output, that profit measures would encourage production efficiency and quality, and that profitable enterprises should be incentivized by increased salary and bonuses. Liberman's proposal initially gained the support of Vasily Nemchinov, a leading Soviet economist-mathematician and early economic cyberneticist, and many others. His ideas also found early favor with Khrushchev, who tested the profit hypothesis in two garment factories. Even after Khrushchev's ouster in 1964, General Secretary Leonid Brezhnev and Premier Aleksei Kosygin, an economic planner committed to systematic reform, continued to support most of Liberman's ideas in the partial and piecemeal roll out of the 1965 Kosygin-Liberman reform.[21]

The fundamental thrust of *Libermanism*, as it became known, was not a sweeping reform of the command economy or its complex accounting (for example, he retains several mandatory target measures in his 1962 article) but rather a retooling and focus of command economy accounting on profit—or what might be called a profit-in-command system.[22] At the heart of these reforms lay an attitude about information that other cyberneticist economists and classical liberal economists on both sides of the cold war recognized at the time: it was an information index that reveals enough about that product and its economic environment to be properly managed. For free-market economists, that golden piece of information was the price of a good; for Liberman, it was the profitability of an enterprise. This reform finds its roots in a compromise between the preservation of the command economy administration and a sideways appeal to ongoing economic calculation debates in Europe. Although Liberman could not explicitly argue against the establishment of a central pricing board (as Friedrich Hayek did in 1945), Liberman's reforms appealed to the efficiency of decentralized economic mechanisms that communicated local knowledge in real time without direct administrative intervention. To Liberman in the late 1950s and early 1960s, it appeared that a self-correcting marketplace of profitability might help eliminate economic inefficiencies, if only factories and enterprises that generated more values than costs could receive their rewards.

Although these two indices—price and profitability—appear to stand as key indicators that distinguish between liberal economists within and without the Soviet Union for streamlining accounting problems besetting any national economy, the opponents to Liberman's reforms insisted that reforming profit measures would also compel a concomitant reform in price: for profit to be a meaningful index, it had to reflect relative scarcities in the economy. This would make visible the hidden subsidies that the state used in the existing pricing system to redistribute resources from one sector to another. It is not clear that Khrushchev understood the full consequences of his decisions: his statements on investment priorities were unclear and changing, perhaps deliberately so, because as a staunch supporter of heavy industry, he enjoyed the discretion to redirect and subsidize certain sectors over others—the very discretion that full profit reforms would have threatened.[23] Nonetheless, the opponents to Libermanism—including the cyberneticists—insisted that, whether in a market or planned economy, all indicators were complexly interconnected. Changing one would surely precipitate a change in the other.

Liberman's reforms met an uncertain end at the hands of those institutions that implemented them in the late 1960s (simultaneously with efforts to advance the OGAS Project for economic reform). Adopted by Aleksei Kosygin and implemented incrementally and partially by a hesitant new general secretary, Leonid Brezhnev, the Liberman reforms nonetheless correlated with increased national production during the next five-year plan (1965–1970), even though they also met fierce resistance from bureaucrats and economic planners, especially in Ministry of Finance, who were set on disrupting the raw materials supply chain and decrying the wage-differentiating reforms as a form of class warfare.[24] By the early 1970s, Brezhnev continued to resist the orthodox economic planners but also abandoned the Liberman reforms.

During the early 1960s, a third camp of thought about national economic reform began to coalesce. The economic cyberneticists championed what might be called *planometrics,* or a combination and application of econometric mathematical tools that included input-output models (not dissimilar from planned supply and demand), linear programming, and sophisticated statistics to the problem of economic planning. Like the liberal reforms, the economic mathematicians, cyberneticists, and econometrists comprising this loose camp conceived of the command economy as a vast information-coordination problem. Unlike the liberal economists, however, the cyberneticists were less concerned with reducing the complexity of the economy understood as an information system to a single golden index. They held that the other two camps did not take seriously enough the numerical nature of all economic exchange and the capacity of modern computing to process them. Mathematicians and theorists such as Leonid Kantorovich, Vasily Nemchinov, Viktor Novozhilov, and B. Mikhalevsky and in the mid-1950s cyberneticists such as Viktor Glushkov and Nikolai Fedorenko realized that universal economic computability meant that all economic relations could be modeled, optimized, and managed with sufficient help from computers and their numerate keepers. In theory, it did not matter which indices were considered, whether price or profit or some proxy variable for peace or propaganda, so long as the boldest socialist ambitions for national economic and social justice could be calculated. In theory, very fast computational speeds made this possible. Computers were thus yoked, quoting Aksel' Berg's book series title on cybernetics, "in the service of Communism" with more enthusiasm than any other toolkit before. By cutting through the political debates of the orthodox and liberal economists, the cybernetists effectively intoned in the face of any economic problem the immortal words of the patron saint of cybernetics, Gottlob Liebniz in 1685: "*calculemus*" or "let us calculate, without further ado, and see who is right."[25]

The most prominent pioneer and precursor to economic cybernetics was Leonid Kantorovich (1912–1986), a prodigious polymath who contributed to the fields of mathematics, economics, and computer architecture. Kantorovich has been compared to John von Neumann (1903–1957), another polymath born of middle-class Jewish parents in early twentieth-century eastern Europe to contribute to the same fields (figure 2.1).[26] Kantorovich's work on computationally optimizing economic exchanges, which later became known as *linear modeling*, began before World War II.[27] The only Soviet economist to be awarded a Nobel Prize (1975), Kantorovich developed linear modeling in 1939 to balance a series of competing variables algorithmically. A simple example adds a dash of dust bowl empiricism to its computational merger of both profit and planning logics. Suppose that farmers—or after Stalin in the 1930s, the managers of a collectivized farm—distribute crops across their fields and that the farmers know the cost of fertilizer and pesticide, the cost of planting, and the selling price of wheat and barley. A linear programmer can determine how much land they should devote to each to optimize their annual yield. Linear modeling—now evolved into the field of *linear programming*—allows the farm managers to calculate in matrix form the maximum revenue, or profit, that they can expect from their available resources and to know how best to distribute their crops (for example, how much barley and how much wheat to plant).[28]

Figure 2.1
Leonid Kantorovich

Economists worldwide recognized the promise of this profit-by-planning model, especially after Kantorovich in 1939 and George Dantzig in 1947 separately took pains to propose methods that could scale to much larger problem sets. Dantzig, for example, showed how the task of distributing seventy jobs to seventy people could be optimized, and Kantorovich's methods found aggregate use in the national wartime efforts to maximize the costs of enemy losses and to minimize those of the Soviet army. (Decades later, their methods remain in use today in modern operations research, such as in Walmart's supply chain.)

Despite the apparent promise of profit by planning in military contexts, linear modeling did not spread in Soviet circles after Stalin had dismissed input-output "balances" as a "numbers game" in 1929.[29] Sped by cybernetics of the late 1950s and the translation into Russian of two articles (a 1958 translation of Wassily Leontief's 1953 edited volume *Studies in the Structure of the American Economy* and an article by Oskar Lange), the majority opposition to economic cybernetic planning methods in 1956 had become a minority position by 1960, and momentum continued to build into the late 1960s.[30] By 1967, the Council on Cybernetics reported over five hundred institutes and tens of thousands of researchers working on cybernetic problems, over half of which featured economic cybernetic research. To this day, the label of *economic cybernetics* lies exclusively within the former Soviet Union and its area of influence.

The scaling successes of economic cybernetics in the late 1950s suggested to Anatoly Kitov, Vasily Nemchinov, Viktor Glushkov, and others that economic planning methods should be applied nationally—perhaps even, as Kitov advised, in a real-time network of computers. The promise of the scalability of the linear programming and computational methods bolstered the political appeal of the supposedly apolitical planometric calculation. The next step with a scalable computational tool is to scale it all the way up, and that would require a communication infrastructure—computer networks—for processing the nation's economic coordination problems. Because computational methods do scale, the economic cyberneticists enthused that maybe *the* principal question for economic reform (who should control the command economy and how?) might be resolved without either the price of politics of the politics of price. It might, the cyberneticists reasoned, be solved with computers.

Many of these proposed reforms—cybernetic, liberal profit, and the Taylorist reforms in the 1920s under Lenin—merited serious attention and, if implemented, would likely have borne fruit had they not collided in application with serious institutional constraints from the bureaucracy. It was self-evident to the economic bureaucracy that computers were not value-neutral: cyberneticists ran them, and no state resolution could convince the bureaucrats to behave like rational bureaucrats in ceding power to cyberneticists. The resulting messy resistance and nonhierarchical dynamics of the administrative base that directed the Soviet command economy reveal institutional tensions and contradictions that foreclosed against multiple attempts to reform the national economy computationally, liberally, and otherwise. Just as Khrushchev's reforms were frustrated and fractured by the internal resistance of administrators who clung to the current positions of power in the late 1950s and early 1960s, so too did the cybernetic appeals to technocratic reform begin to break against the practical problem of reforming a national economy that refused to behave like the hierarchical system that it appeared to be on paper.

Liberal economists and economic cyberneticists (at least initially, under Viktor Nemchniov) appeared to be bound for a great alliance. In the early 1960s, Nemchinov proposed a "self-supporting (self-accounting) system of planning" that integrated both decentralized computational and market mechanisms into the planning apparatus. The basic proposal was to solve the incentive problem in a way that no factory would have a reason to act against the wishes of the center and the center would have no reason to compel the factory to act.[31] With time, however, the cybernetic economists and the liberal economists clashed over whose method would win the balance of state approval. In 1963, both Liberman's profit proposal and Glushkov's OGAS project appeared positioned to affect real economic reforms. Leading liberal economists, including Evsei G. Liberman, A. M. Birman, and B. D. Belkin, voiced the public opposition to mathematical economic reform in general and the OGAS Project, in particular, although without spelling out the secret project by name in the press. These leading liberal economists immigrated to the United States and Israel after the Liberman-Kosygin reforms were formally accepted but botched (or rather deliberately butchered) by the administrative apparatus. Birman criticized the economic cyberneticists not for their methods but for their politics. As late as 1978, he contended that the introduction of computers and automated systems of management (ASUs) into Soviet economics constituted no more than a "costly delusion" and was a question of the complexities of human interests, not precise accounting.[32] In effect, the liberal economists accused computational economics of harboring conservative politics and of trying to work in the framework of the existing political system without any social change.

Most cyberneticists came from the technical, theoretical, and natural sciences—fields that attracted many of the brightest Soviet minds because of the state support they received and the safety of choosing ostensibly nonpolitical specialties. The competition to join the top sciences was immense, and few chose to leave the sciences for the social and humane sciences (Kitov was forced into economics, and Glushkov was an exception). Leading figures in the mathematical economic camp (such as Kantorovich, who received the Nobel Prize in 1975) were known to defend orthodox political values about the price of labor, even while the younger generation of cyberneticists sought to avoid the politics of price by arguing that a sufficient change in the organizational values of the system must also cause a concomitant change in the political values. By attempting to rationalize and decentralize the planning process, the cyberneticists hoped that anyone, with the help of a computer, could contribute to a reformed, well-oiled economic model and plan, make the system work better, and open a quiet back door to political reform. Even so, Birman and other veteran economic reformers wondered whether the deliberate planning that was inherent in a cybernetic reorganization of economic planning would exacerbate and reaffirm preexisting constraints and coordination problems in the command economy. The liberal economists saw in cybernetic reform of the planning administration no promise of a transition to the market economy that they sought. This belief that technological and organizational reforms bring political ramifications recurs as an article of faith in the annals of Soviet cybernetics.

Despite Glushkov's complaints to the contrary, it is not clear that liberal economic opposition to the economic cybernetic school held up or delayed Soviet attempts to carry out economic reform by computer networks.[33] By 1970, when the top echelons of the Party were ready to consider such proposals in earnest, the liberal economic opposition to the cyberneticists might have helped ingratiated the cybernetic cause to more orthodox Party members who were fed up with Libermanism. (By that point, liberal reforms had a five-year track record of generating more heat than light in many of the economic administrations.) At the same time, the military and the Party were tantalized by the promise of a third generation of integrated circuits in computing in 1970s and maybe even the fourth generation of microprocessors on the horizon of computing industries abroad. Given the political and technological climate, the ears of the state were primed to hear Glushkov's declaration that "the scientific-technical revolution has thrown such a challenge to the science of governance, and much will depend on how we dare to answer that challenge."[34]

Vertical Bargaining and Other Organizational Dissonance in the Soviet Command Economy

This chapter's consideration of the inner workings of the command economy looks directly into the political heart of socialist economic reform, of which the cybernetworks were a small part. To organize an economy properly was the litmus test for the Soviet experiment in socialism and social justice. Few other national projects claimed and endured as much in search of a Soviet network. As a consequence, the organizational dissonance that all economic reformers, not only cyberneticists, encountered in trying to make the economic numbers line up was both the cause and effect of continual economic reform. In this industrialist mindset, computers brought to perennial problems a new set of tools (linear processing, input-output modeling, and the possibility of real-time network communication and surveillance). This section examines some sources of what David Stark has called the "organizational dissonance" that underlies the command economy and that helped ensured the economic system could not be reformed or reaffirmed because every reform introduced new problems without solutions.[35]

"Vertical bargaining" was a feature, not a bug, of the perpetual misalignment of incentives in the Soviet economic hierarchy. Named by a Hungarian economist and critic of socialist economic systems, János Kornai, vertical bargaining takes place among the three levels of relationships among a local enterprise, a branch directorate, and the national planning ministry. Vertical bargaining took place continuously in the annual planning process that, as Spufford describes it, pulsed with paperwork between Gosplan, regional councils (or Sovnarkhozy between 1957 and 1965), and the enterprises (such as firms, factories, farms). Every spring, the enterprises asked Gosplan for the supplies they needed as a percentage change from the output of the previous year. Around the end of June, Gosplan sent draft production targets to the regional councils, which disaggregated the targets and then negotiated with the enterprises toward trim but not unmanageable requests for inputs. Gosplan then reaggregated these requests into each commodity's total supply for the nation that year. When the figures did not match, a second negotiation period between Gosplan and the regional councils proceeded into the autumn until Gosplan had limited demand and maximized supply. The finalized supply quotas and production targets could then be passed down the chain in late October to allow enterprises to select next year's items from the "specified classification," a list of every item that officially was produced in the Soviet Union (think of the Sears mail order catalog on steroids, minus the advertising), just in time for the process to begin again.[36]

All negotiation processes were structured for expressing "mutually contradictory motives," although the administrators had no special access to mechanisms for resolving a priori conflicts between the interests of the layers of the hierarchy, which compromised the integrity of the economic plan that they were developing.[37] Without a plan for regulating the planners, the planning processes confronted economic leaders—everyone from administrative planners to factory managers—with multiple registers of conflicting value. "Suppose a leader feels he has received an incorrect order," Kornai asks: "Should he carry it out or should he protest, out of party loyalty and professional pride?"[38] If he accepts the flawed order but fails to deliver on it, Kornai continues, he and his colleagues will be held responsible and possibly accused of sabotage. If he opposes the order, he could be accused of party disloyalty. Either way, the actor, not unlike Vanek in Vaclav Havel's play *Audience*, is stuck.

Without a single path forward, levels had to negotiate for their own institutional self-interests vertically across the formal administrative hierarchy. To do so, requests began to misrepresent economic reality in both directions. Requests for input (or demand) rose upward and request for outputs (or supply) sank downward—the planner's vertical equivalent of selling high and buying low in a horizontal market. Imagine the behavior of the ministry that oversees a branch directorate and the factories that the directorate oversees. The branch directorate is charged with reporting to the ministry statistics about the annual production, material allocation, and labor of its subordinate factories. To do so and because the experienced directorate anticipates that factory managers are responsible for shortfalls and thus systematically underestimate their output capacity and overestimate their input needs, the director will "prescript a plan 10 or 20 percent tighter than they themselves consider realistic, calculating that the firm will want to beat them down."[39] When reporting its plan, the branch directorate bids to the superior ministry just as the factory manager did to it. An apocryphal anecdote of a job interview for a new accountant in a factory captures something of this haggling spirit. To each candidate, the factory manager asks only one question: "How much is two and two?" A single candidate, a former convict, has the winning answer: after hearing the question, he stands, closes the door, and asks in a loud whisper, "How many do you need?"

The vertical bargaining process also penalized the future of productive factories by "planning in" their previous successes as the new baseline, ensuring that the plan would be ratcheted upward in perpetuity.[40] Manuel Castells notes that the entrepreneurial managers and workers in the chemical complex of Shchenkino in Tula, Russia, "were trapped into being, in fact, punished with an intensification of their work pace while firms that had kept a steady, customary level of production were left alone in their bureaucratic routine."[41] This ratcheting effect, a kind of institutional-ized variant of the tall poppy syndrome, has its corollaries in the mutu-ally reinforcing relationship between demand and supply. In corporate and command regimes, if the supply of one's quality goods meets demand, one must work harder to meet future elevated demand. If the supply of one's goods falls short of the need, the capitalist market actor will adjust or go bankrupt, and the socialist administrative actor will be punished. So long as labor is isolated from those who manage the means of production—Marx himself railed against the doyens of exchange value (*Tauschwert*)—manage-ment profits and alternately pays or punishes workers for past productivity. Unlike market behavior, any deviations from the plan could send culpable ripple effects down or up the chain, and the plan itself could be understood in the context of its local knowledge. So the ideal standard of factory or firm behavior is to fulfill the plan by "exactly 100 percent, or perhaps 101 or 102 percent." The lively fitfulness of vertical bargaining disrupted and distorted the representativeness of the economic statistics that the cyber-neticists sought to input into the linear modeling and programming and economic reform projects.

Shoehorned instead into numerical fit, the command economy plan never fully reconciled in its manifold details, and the compounding compli-cations and activities that organizational dissonance invited were anything but planned. Administrative roles blurred as needed. One leader might find himself playing the politician, a bureaucrat, a technocrat, and a manager in different situations. To secure more input on the accounting sheets, a leader might encourage workers on the factory floor to produce more out-put in the name of the plan but, behind closed doors, might misrepresent the numbers to undercut the same plan. Not at all the rigid hierarchy cari-catured in modern memory, the everyday practices of Soviet economic life abounded in pervasive informal forms of competition and caprice.

These early cybernetic economists faced an associated and monumental challenge. In theory, their reforms would be approved to track the formal (or white or first) economy, and any attempt to reform the distortions in that formal economy would further incentivize more informal (or gray, sec-ond, or shadow) economic activity. The reality of the latter half of Soviet economic history reveals that private life increasingly depended on the resilience and robustness of people's connections to the informal economy.

The command economy operated on hidden networks of *tolkachy* (literally "pushers") or "go-to-guys" or "fixers" who got the job done outside of the formal economic plan. Without the support of *tolkachy*, thousands of official economic quotas over decades never would have been met.[42] Recent economic studies based on previously unconsulted archival data have estimated that a stunning average of 24 percent of annual expenditures per household in the Soviet Union went to the informal economy between 1968 and 1990.[43] The estimated percentage of GNP not accounted for by the informal economy over the same period ranges between 17 and 40 percent.[44] Despite both official claims and its enemies' fears otherwise, Soviet economic life drew its vitality not from the strictures of top-down command and control but from the fitful hustling and the scrambling that came about because of those commands.

Informal behavior and bargaining were not separate from Soviet statecraft: the state embodied them. Even Stalin, with his reputation as an all-knowing leader and steely strongman, bowed to the deeper logic of informal influence and favors—what is known simply as *blat* (ostensibly from Polish Yiddish for "someone who covers for someone else," or from the German for "blank note").[45] Instead of committee decisions, Stalin often invited local leaders to private consultations where Stalin could claim that all other parties had endorsed his recommended policy and provide the local delegates with an opportunity to leverage "personal connections" to personal advantage at home.[46] Control over science and society was extended by the same informal means. In 1952, an editorial in Hungary proclaimed that "the teaching of Stalin embraces all the universal principles of nature in its smallest details. He solves all the practical problems of understanding natural science," and "it is only Stalin … who is able to analyze clearly and find with mathematical precision the exact way toward solution of present day problems."[47] So too did his social radar appear impossibly omniscient thanks to strategically placed ambassadors and Party secret police embedded with party *apparatchiks*. The strongman seemingly did not want the state to behave as a well-ordered hierarchy but rather as a sprawling network with informal connections to a strong but sporadic center. The terror of his rule was not its rigid centrality but its informal uncertainty. Stalin and his henchmen could be anywhere. To encompass everything was Stalin's job: he already had everything covered. And for this reason, his death formed the vacuum into which cybernetics stepped.

Khrushchev's thaw attempted to distance the nation from its Stalinist past, but his decentralizing reforms sped the sporadic, informal, and unregulated nature of his rule. Even his famous "secret speech" to a crowded Party congress in 1956, which was the act that distanced him from Stalin, denounced Stalin as a cultish personality but *not* as a governor whose mode of informal management Khrushchev wanted to break from.

The administrative infrastructure came to reflect this infusion of informal administration in a number of ways. Administrative personnel and staff officials had separate telephone lines and mailboxes for the same supervisor, which ensured that formal communication lines were clogged with official requests and that the actual negotiations took place along informal lines—not on the golf courses of modern business but in the transit sites such as hallways, trains, and dachas (seasonal cottages or summer homes outside the city). Because formal mechanisms proved ineffective, hiring and promotion practices often relied on interpersonal and informal "career friendships" or tight bonds that lasted lifetimes. Soviet specialist David Granick notes that "with this absence of formal clarity, it is natural that emphasis has always been placed on the need for the closest ties and a comradely atmosphere between the management and a plant's Party organization."[48] Interviews with émigré bureaucrats have revealed a pattern of administrative behavior that stressed the career necessity of not "spoiling relations," the significance of who you know, and the career advantages of being a "yes-man" in formal relations with superiors.[49] Administrative conflicts between the elements of that system—such as the Academy of Sciences and the Ministries of Finance, State Planning, Interior, Defense, and State Security (home of the KGB, or Committee for State Security)—were resolved not by an appeal to hierarchical authority but through a variety of informal mechanisms that were internal to the ministries themselves. One reform initiative after another was aborted, and those that were enacted were condemned to stumble on, in Schroeder's phrase, the late Soviet "treadmill of reforms."[50]

Compelled to operate within an official hierarchy, Soviet administrators benefited by behaving and working across complex informal networks that crisscrossed across institutional interests. A young economist, Menshikov, summarized the nonlinear or nonhierarchical behavior of the command economy not as "aiming at increasing the well-being of the population" but as "maximizing the power of the ministries in their struggle to divide up the excessively centralized material, financial, labour, natural, and intellectual resources."[51] He continues, noting the path-dependent creep of administrative misbehavior: "Our economic-mathematical analysis showed that the system had an inexorable inertia of its own and was bound to grow more and more inefficient." Whether in vertical bargaining between the levels of the planning, the reluctance of ministries to cooperate outside their assigned territory, or the struggle of other agencies to collaborate, the fundamental paradox of planning became increasingly obvious to the economic cyberneticists and others who sought systemwide reform. The planners and executers of that plan were to create gaps in the plan and leverage those gaps with very competitive logics that the command economy sought to prevent.

The cyberneticists thus faced a foundational paradox in reforming the national economy and perhaps other political economic systems. For an economic reform project to succeed politically at the national level, the reformers had to win the support of the very system that it meant to reform. If they sought to do so through new formal mechanisms, as the computational methods of the economic cyberneticists demanded in theory, those methods would face widespread resistance (one of few systematic behaviors that the system was regularly capable of). Conversely, if they sought to reform a broken system of political patronage, as they had to do in practice, they first had to win the favor of that system. The paradox that they faced is not unique to the Soviet cyberneticists. To reform a system, a would-be reformer first has to become part of it. Next, the better that one plays along, the less likely that one wants to reform the system; and so long as one continues to play along, one may not reform the system.

Faced with technocratic reform, economic management bureaucrats and politicians scrambled their own administrative orders to preserve their own personal careers. Bureaucrats were never mere bureaucrats, and the mechanics of day-to-day operations were never merely mechanical, even though the culture of technocratic governance swelled after Stalin to the point that, by 1989, 89 percent of those who sat on the Politburo were trained engineers (engineering training prepared Soviets for governance positions much like law degrees do elsewhere).[52] (The iconoclast economist Thorsten Veblen mused in 1921 that the West might one day be ruled by a "soviet of technicians" or a technical class that was capable of capturing the wealth that they produced.[53]) The Soviet system, much like a firm, sought to produce one solitary good above all else—the political good of a life apart from the capitalist experience. In a narrow sense, it succeeded: the marketplace of Soviet economic interactions became foremost a negotiation of political power rather than price. Its bureaucrats bowed to unintended incentives to exploit the rampant organizational dissonance that they oversaw, its technocrats lived by their social wits, and the system squeaked by on the capricious politics of planning run amok.

Conclusion

As Soviet economic cyberneticists emerged as a viable option in the early 1960s and again early 1970s, they confronted a monumental problem in managing and reforming the command economy. The question was identified by the Austrian school of economics decades earlier: which techniques and approaches would help resolve the mounting tensions among the formal command economy, the gray economy, and the infusion of informal practices in the administration of Soviet socioeconomic life? For the most part, the leading Soviet economic cyberneticists sought to fix the formal command economy by introducing ambitious, even grandiose, plans for automating and modeling the administrative planning decision process itself. And yet, as is shown, those formal plans—a networked plan to fix the planning process itself—did not work because even cybernetic plans could not account for nonlinear operations in the Soviet economy. Their formal plans to rebuild the command economy as a hierarchy had to overlook the complex crisscrossing networks of relations that made it function in practice—the gray economy and its entrenched currency of *blat* or informal favors that were entrenched in the governance structures. By reimagining the command economy as a heterarchical crisscross of hierarchical orders from above and a resulting swirl of unregulated practices in every other direction, the failure of these Soviet economic cyberneticists to reform, automate, and manage the command economy begins to make more sense.

There is one goliath exception to this critical description of the informal administration of the state and economic bureaucracy. The command economy, which staggered along a winding path toward the creation of a normal industrial civilian economy, was relatively functional at powering and sustaining superpower military technological initiatives. Formulated first as a wartime economic model by the Germans, the insatiable sink of the Soviet defense apparatus into which economic resources were continuously poured cannot be overestimated. Both official state and CIA statistics on Soviet military spending are controversial, although if a critic of both can believed, the Russian American economist Igor Birman estimated that by 1975 the CIA estimates of the size of the Soviet economy were two or three times larger than reality and that instead of spending roughly 6 percent of its GDP on military expenditures, the Soviet state devoted closer to 30 percent of its GDP on the military. The military-industrial complex enjoyed massive funding streams and the brightest and best intellectual and technological resources, and although the jury is still out on the exact nature of the Soviet military (most of its details remain closed to this day), the military sectors also attracted the best and the brightest because those sectors were best managed. The strictly managed military sectors produced and sustained for decades world-class space and nuclear programs and secret computer networks across launch pads deep into Siberia. But the Soviet military's technological innovations did not as a rule spill over into civilian sectors. Nuclear-blast-resistant computer chips interested very few, and yet the Soviet military nonetheless developed these and other "special research" projects (so named in public documents).[54] The military enjoyed the status of being responsible for generating a seemingly infinitely defensible "public good"—national defensive and offensive readiness in an almost irrationally strategic cold war—and yet did not have the burden of having to be publicly accountable to civilian politicians.[55] Perhaps the caricature of the problems of the civilian economy makes most sense in light of its foil in the military economy. Unruly, informal, labyrinthine, and ineffectual suffering in the civilian sectors rarely met with the well-ordered, formal, hierarchical modernization in military affairs. The contrast between military and civilian economies recapitulates a structurally similar disconnect between the civilian command economy in theory (hierarchical, formal, well ordered) and in practice (heterarchical, informal, conflict ridden).

In summary, the separation between military and civilian sectors reflected a disconnect between the civilian economy and its own state goals, and it comes squarely into play in the central story, outlined in the following chapters, about early Soviet computer networks projects and those cybernetic entrepreneurs who set out to build them. This chapter has laid out the basic civilian economic operations as well as problems that motivated Soviet cyberneticists—whether orthodox, liberal, or cybernetic—to propose and design ambitious projects for reforming the national economy. The next two chapters examine the role that computer networks played as the promised deliverers of such reform, and they turn on why cybernetic attempts to network the command economy fell apart. The network entrepreneurs understood firsthand the institutional contradictions that they sought to solve with an automated system of management. This basic backdrop to the everyday administrative conflicts in Soviet social life—between the rational hierarchical plan of the command economy and its messy heterarchical misbehavior—was not lost on the pioneering Soviet cyberneticists who followed.

3   From Network to Patchwork: Three Pioneering Network Projects That Didn't, 1959 to 1962

In the late 1950s and early 1960s, economic cybernetics—with its nonlinear mathematical mindset—appeared to be a near-perfect approach for modeling and reforming the economy's heterarchical coordination problems. (Cyberneticists were at home in the underlying observer-effect problems: the act of looking at a system changes the system.) Between 1959 and 1982, Soviet cyberneticists advanced a half dozen ambitious and creative plans to network their nation, including four overlapping attempts to digitize the command economy. Most of those proposals arose between 1959 and 1962 before they languished or merged with associated projects with more momentum, like the OGAS Project described in chapters 4 and 5.

This chapter reviews the three earliest Soviet network ambitions—Anatoly Kitov's EASU (Economic Automatic Management System), Aleksandr Kharkevich's ESS (Unified Communication System), and N. I. Kovalev's rational system of economic control between 1959 and 1962—in the context of the larger institutional struggles to secure support for their network projects. Some attention also is paid to the network projects that developed as civilian computer networks elsewhere and to the ways that the Soviet experience varies from the American ARPANET and Chilean Cybersyn projects. Through a discussion of leading Soviet network proposals to reform the command economy in the last few years of Khrushchev's reign (1959–1962), this chapter examines, details, and complicates the hypothesis that the administrative dynamics of a strong civilian-military separation help explain the stillbirth of their historic efforts.

Anatoly Kitov and EASU: The First Soviet Cyberneticist and His Civilian-Use Military Network

The first person to propose a large-scale computer network for civilian use anywhere, as far as I can tell, appears to have been the first Soviet cyberneticist, Anatoly Ivanovich Kitov (1920–2005). The son of a White army (Menshevik) officer who escaped persecution after the 1917 Russian revolution by moving from Moscow to central Asia and then to the city of Kyibishev (now Samara, Russia) on the Volga River, Anatoly Kitov grew up between two world wars. A star student in mathematics who rose rapidly through the military academy, Kitov served as a young officer on the front in World War II (where he, like other human "computers," computed ballistic tables), before launching a distinguished military career that suddenly shifted to civilian research for reasons described below (figure 3.1).

In 1953, Aksel' Berg, then deputy minister of defense in charge of radar and future dean of Soviet cybernetics, asked Kitov to prepare a report on the state of computing in the West.[1] Kitov's optimistic report resulted in the creation of three large computational facilities—the Computation Center 1 (which Kitov directed until 1959), the Navy Computation Center, and the Air Force Computation Center.[2] Kitov's optimistic review of computing in the West stemmed from his 1952 discovery of a copy of Norbert Wiener's *Cybernetics* that had been removed from general circulation (due to the ongoing anti-American campaign against cybernetics) and stored in a top-secret military research library. As noted above, in 1955, Kitov coauthored (with Lyapunov and Sobolev, two highly regarded Soviet mathematicians) the first Soviet article to attempt to rehabilitate cybernetics from

Figure 3.1
Anatoly Kitov. Courtesy of Vladimir Kitov.

the anti-American ideological critique that had been waged since its first mention in the Soviet press in 1948.

Kitov was not alone in seeing the potential for using computers in military work. Military and computing innovations were inseparable in the early history of computing. Although those early, specialized computer innovations for the military often had no measurable defense outcomes, their technological innovations seeped into nonmilitary industries. (Examples of military products that are now available commercially include jet planes, semiconductors, telecommunication and computer equipment, microelectronics, sensors, GPS, drones, and even Velcro, and only a few consumer electronics, such as game consoles and consumer electronics, have run the other way.[3]) In the United States during World War II, early military computer projects included the Whirlwind I (a vacuum tube computer), the Whirlwind II, and early attempts at computerized command and control, which in the 1950s led to SAGE (Semi-Automatic Ground Environment), which was a radio and radar network that stretched over most of Canada and was intended to intercept invading bombers from the Soviet Union. For its military purposes, SAGE was obsolete before it was operational, but its preparation nonetheless sparked a wave of influential inventions and major technical advances in computer systems and networks, including magnetic core memory, video displays, graphic display techniques, analog-to-digital and digital-to-analog conversion, multiprocessing, and automatic data exchange among computers.[4]

Alarmed by the news of SAGE in the mid-1950s, the Soviet military responded with at least three major long-distance computer networks—a missile defense system (System A in the late 1950s), an air defense system (the TETIVA in the early 1960s), and a space surveillance system (beginning in 1962). In the late 1950s, for example, a prototype missile defense system that was code-named System A (about which little else is known today) was built around a computer network that connected two Soviet mainframes, the M-40 and M-50, and a series of specialized computers at remote radar installations. Soon after the successful testing of System A in March 1961, Khrushchev boasted that Soviet antiballistic missiles could, in his famous phrase, "hit a fly in outer space."[5] More significant than the system's accuracy, however, was the fact that System A and its sibling military networks compelled a larger geostrategic shift: namely, antiballistic missiles that pointed skyward around the world greatly diminished the strategic value of a first-strike attack. Other networks, such as the space surveillance system started in 1962, connected a pair of distant nodes (one near Irkutsk on Lake Baikal in eastern Siberia and the other in Sary-Shagan in south central Kazakhstan) to a master computer center just outside of Moscow some seven thousand kilometers away. Other radar networks—named Hen House, Dog House, and Cat House—bathed vast swaths of territory in anti-ballistic alerts. The Dead Hand—the semiautomatic perimeter defense system noted in the introduction—is the present-day heir to these early Soviet military computer networks.[6]

Struck by the latent computational surplus that was available in military networks, Kitov turned his attention to how networked computing might be able to benefit the civilians who were supposed to be protected by the military.[7] In 1956, in the first Soviet book on computers, *Digital Computing Machines*, Kitov built on the insights of Leonid Kantorovich's linear modeling to promote a pioneering argument about the potential of the computer—although still without the network—as an essential tool for modeling, programming, and regulating the Soviet economy.[8] The idea of using high-speed digital computers to crunch economic statistics was nothing new. In fact, Kitov's proposal came as a mere technological update to a long-standing tradition of state-based mechanical computation. After the 1917 revolution, for example, the Bolsheviks quickly nationalized the Odhner calculator factory in St. Petersburg. By 1929, the year that farms were mass collectivized and planned, the Soviet Union employed statistical tabulation equipment—including a clone of the successful pinwheel Odhner arithmometer and IBM punch card machines—on the scale of the United States or Germany.[9] The etymology of the word *statistics* (German for "the science of the state," which replaced "political arithmetic" in English in the late eighteenth century) frames this trend as nothing peculiarly Soviet: modern states, from the census to taxation and conscription, have long made statistics their business.[10]

The initial idea of using unnetworked computers to process economic information—really no more than a modest technological upgrade that was in line with well-established state interests—had been gathering support since at least 1954, when Mikhail Kartsev, an engineer who participated in the construction of the M-1 computer, declared that cybernetic understanding of computers had to exceed narrow military tasks: "we are interested not so much in the military applications of mathematical machines or, more generally, new technical devices, but in their wider applications." His colleague Nikolai Matiukhin, citing the use of computers in U.S. business, stressed that "in a socialist country, ... the mechanization of planning with the assistance of computers can and should be pursued to the largest extent possible."[11] The early Soviet information technologist Isaak Bruk, who developed the M-2 computer, picked up on the thread, publishing the first call to harness computational power into raising the quality of economic planning in a 1957 *Kommunist* article, "Electronic Calculating Machines: In the Service of the National Economy."[12] (Both Kartsev and Matiukhin continued serving in military careers.)

Not long after, from 1958 to 1959, Kitov and Bruk's proposal began to bear fruit as Gosplan constructed specialized computational centers (*vyichislite'nyi tsentri*) for economic accounting, which were to be under the control of the Economic Council within the Council of Ministers. Vasily Nemchinov, a leading Soviet mathematical economist, also championed the proposal for computer centers around the nation to improve planning. In January 1959, Kitov sent his first proposal directly to the top of Soviet power and called on General (then First) Secretary Nikita Khrushchev to recognize the need to use computers to process economic planning and to speed economic reforms, and with the letter he included a copy of his published book on digital computers.[13] None of these proposals mentioned a computer *network*—or its near synonyms (such as *base*, *system*, or *complex*)—that would be able to command the economy.

Kitov's first letter to the Central Committee in 1959 proved to be a success. Although Khrushchev probably never saw the letter, his message ended up in the hands of Leonid Brezhnev, who replaced Khrushchev as general secretary in 1964. A trained technologist who had studied metallurgical engineering in eastern Ukraine, Brezhnev approved the proposal and ordered a government resolution to review and execute Kitov's recommendations. A commission led by Aksel' Ivar Berg was created to enact the Communist Party resolution called "The Speeding and Widening of the Production of Calculation Machines and Their Application to the National Economy." Berg cut a cosmopolitan figure (Berg's mother was Italian, and he was a Swedish Finn) in the administrative support of Soviet cybernetics, and he rose through the military ranks to serve on the State Committee of Defense after World War II with special emphasis on military and naval matters. Rescued from a sunken submarine near Helsinki in 1918, he directed radio technology research for the Red Navy, was imprisoned during the Great Terror for spying, returned to research during World War II, and was vice minister of defense in charge of national radar and radio technology from 1953 and 1957. In this position, Berg provided critical administrative support to the efforts of many early cyberneticists, including Kitov, Lyapunov, and others. Appointed to serve as the chair of the Council for Cybernetics in 1959, Berg, perhaps not unlike Vannevar Bush, exerted an extensive range of administrative influence over the state of Soviet cybernetic science for the next two decades.[14] With Berg's support, Kitov's first letter to Khrushchev set into motion a sea change that swept up strong state support for cybernetics.

A signal political victory, Kitov's initial vision imagined computers as devices for local computation but not yet for national communication. Like his predecessors, he proposed that "electronic calculating machines" must be used in "automating administrative and economic governance" in planning for the then seven-year plan for the command economy.[15] He also called for a "reduction of the administrative-management personnel" for engaging in "outdated means and methods of leadership." This could take place at local levels by means of local control systems, called automated systems of management (ASUs) (*avtomatizirovannie sistemi upravleniya*). ASUs were automated control systems at the level of the factory—a kind of local area network that allowed mainframe computers to control and communicate with factory machinery through a series of automated feedback loops and programmable control processes. In the 1960s, ASUs were developed and implemented in individual Soviet factories incrementally, and their use increased slowly in the 1970s and 1980s.[16] By the early 1960s, the notion of ASUs—or computer systems for monitoring local industrial processes and automatically optimizing those processes for efficient outcomes—gained popular traction among factory and enterprise managers around the country. On the cusp of the 1960s, enterprising military researchers like Kitov saw the advanced computer technology behind ASUs as promising new efficiencies, savings, and economies of scales at the factory and enterprise levels.

The next step was to nationalize the ASU and make it go "all-state" (adding the prefix *OG* for *obshche-gosudarstvennaya* to form the OGASU). This step appears obvious today, but the consequences of that step must have been hard to foresee then. In fact, encouraged by the success of his first letter, Kitov shifted his attention from local computation to national communication. In the fall of 1959, he drafted and sent to the Party leadership a second, more ambitious letter that eventually became known as "the Red Book" letter due to the color of its cover. The Red Book letter embraced a far more radical idea—the first wide-area dual-purpose computer network that could support both military and civilian uses. In his proposal, Kitov conceived of a "unified automated computer network" for administrative control of both military and economic affairs that would be built on the extant territorywide lattice of Ministry of Defense computer centers. Although electrical telegraphs had been instantaneously connecting paying publics across long distances since the mid-nineteenth century, evidence indicates that Kitov's 1959 proposal for a dual-use network was the first anywhere to suggest allowing civilians to use military computer networks to work on national problems.

Kitov named this first national network the Economic Automatic Management Systems (EASU, for *Ekonomicheskaya avtomatizirovannaya systema upravleniya*), a short step from the local area networks of the ASU. The EASU was meant to be more than a colossally oversized ASU, or factory-based automated management system, however: it was to be a dual-use network that would overlay and manage existing information flows within the Soviet economy with "large complexes" of computer centers. The long-distance economic ASU began to articulate in technological terms the underlying Marxist conception of the national economy as a single complex industrial body. The informing values for the earliest Soviet networked vision were in context both technologically ambitious and political self-evident.

The EASU proposed for the first time a long-distance communication infrastructure to transform the command economy into what it had effectively thought itself to be—a single nationwide corporation devoted to producing one product, which was social life outside the reach of capitalism. The basic communications infrastructure for such an upgrade was fairly straightforward. What Kitov called a "complex of computers," or a computer network, would use Ministry of Defense computers to optimize national economic planning and streamline the bulky and inefficient administration for planning the Soviet national economy. Each powerful computer center in the network would build on military computing locations that already were underground, well protected from the threat of enemy bombing and natural interference above ground. In addition, each underground military computer center would connect to accessible computer terminals that were located in cities above ground where "civilian organizations" could receive, send, and employ "unlimited quantities of reliable calculating processing power."[17] The military's automated missile computer network systems would serve, in Kitov's vision, as the technical platform for computationally monitoring and managing the national economy.

Kitov went considerably further than most military men of the time in proposing dramatic financial savings and benefits for the state. Electronic economic reform would also help quicken, Kitov added, the currently sluggish and inadequate adoption of computer technology by the Ministry of Defense. As part of that criticism, he called for the creation of a new governmental body that would be charged with overseeing the reform of all institutions, including both military and civilian, that were associated with planning the national economy.

With a proposal that criticized the military and proposed a civilian-military project, Kitov sent his letter to Khrushchev sometime in the fall of 1959 with the "hope that it would be accepted," he later remarked, "just as easily as the previous one."[18] Unfortunately, the precise fate of Kitov's second letter—the Red Book letter—remains unclear (in 1985, Kitov recalled that his letters to Party leadership were "jammed," or *strevali*).[19] We do know that the second letter never arrived at the desk of Khrushchev, Brezhnev, or any member of the first secretary's inner circle like the first letter had. Instead, the letter—with its criticism of the military and proposal to share military technology with civilians—fell into the hands of his military supervisors, who were infuriated. Without Berg's support and under his military supervisors' initiative, a special military commission was convened to review Kitov's report.

The highly respected high-ranking war hero Field Marshall K. K. Rokossovsky chaired the commission as then chief inspector of the Ministry of Defense. Rokossovsky—who survived the great purge, show trials, and torture under Stalin—might have been sympathetic to Kitov's case had he actually attended the commission. As it happened, however, Rokossovsky barely participated in the commission, leaving Kitov's fate in the hands of his supervisors, who rejected the proposal and followed standard Soviet procedure in burning the unapproved (and irreproducible) proposal in what colleagues later referred to as Kitov's show trial.[20] "Hence the paradox in technics," as Lewis Mumford put it: "war stimulates invention, but the army resists it!"[21]

Incensed by Kitov's critique, the unchecked commission exacted further retribution by revoking his Communist Party membership for the following year and dismissing him from military leadership, his position as the director of Computational Center-1 of the Ministry of Defense, and effectively his once meteoric military career. To justify this punishment, the special commission deemed Kitov's proposal "inefficient" for having suggested that civilians should use of military technologies, disregarding any discussion of his promised cost savings and efficiencies. The commission also issued a formal complaint against Kitov for not having filed his network proposal according to proper protocol. He was to be punished formally for having attempted to send his communiqué to Khrushchev directly, bypassing the intervening administrative tiers between him and the Party leaders.

Given the success of his first improperly filed letter to Khrushchev earlier that year, a breach in filing protocol struck his colleagues as a disingenuous and insufficient cover for severely punishing an army researcher who was celebrated for having done something similar a year earlier. Eyewitnesses confirm that the commission's unwritten response had little to do with filing protocol, efficiency, or any other stated reason. Instead, they underscored the military's possessiveness and unwillingness to share information technology with the civilian sector. As Kitov later exclaimed in an interview, the commission's unwritten response was essentially that "the army will never occupy itself with fulfilling any tasks concerned with the national economy!"[22]

Most notable about Kitov's show trial is not the possessive self-interest that motivated a large institution to punish its own—an instinct that animates most centralized command-and-control administrations, including the military portion of the Soviet knowledge base—but rather the counter-innovational institutional conditions that sanctioned the military command to separate military and civilian resources, both economic and technological. The military top brass decided to hoard its computing resources, neither acknowledging the Politburo's support of Kitov's proposals nor concerning itself with any commercial or civilian application. Kitov's military supervisors were free to act as they pleased, denouncing any time sharing of their computer networks with others, even if doing so at night would have had no obvious cost to the military and would work against the interests of the top state leaders of the nation that the military was sworn to protect.[23]

Kitov's show trial also showcases the informal and contingent dynamics that beset anyone trying to bridge the entrenched military-civilian divide. Although Kitov's first letter circumvented formal protocol without a hitch, the second letter, which included military criticism, was intercepted. The military man who was most likely to be sympathetic to Kitov's case did not attend the commission that he formally chaired. Informal degrees of freedom, in turn, allowed Kitov's military supervisors, according to eyewitness reports, to pronounce his proposed military-civilian nationwide network an existential threat—not as much to the nation as to their personal and unprecedented control over the resources of the nation.[24] An automated computer network threatened to automate and jeopardize the Ministry of Defense's positions of power over strategic bottlenecks of resources in information technology, granting civilian economic planners access to the ministry's technological monopoly.

Kitov's first public computer network proposal ended his military career and launched his career as a civilian network entrepreneur. In his many publications promoting automated computer networks in the national economy between 1959 and 1967, he continued to frame the economic race in terms of military competition between the superpowers.[25] In a 1959 article with Berg and Lyapunov, for example, Kitov announced to his readers that the automation of firm-level economic management resulted in major savings and "reductions in the administrative apparatus (in some cases 80–90%)."[26] In 1961, his advice to reform economics with computer methods—without the military network—successfully secured the support of the top Party leadership in the form of a Party report that helped pave the way for Khrushchev at a Party Central Committee Plenum in November 1962. At that plenum, at the height of the cultural thaw and the eve of the economic debates described previously, Khrushchev called for the adoption of Western "rational" managerial techniques, proclaiming that "in our time, the time of the atom, electronics, cybernetics, automation, and assembly lines, what is needed is clarity, ideal coordination and organization of all links in the social system both in material production and in spiritual life."[27] In many ways, influencing the leaders of the Soviet state with cybernetic ambitions about networking the civilian economy proved easier than bridging the military-civilian divide.

The Soviet military behaved as a well-oiled hierarchy when it limited scientific or technological transfer outside of itself, although like the economy and Party apparatuses, its internal affairs could be unpredictable and tenuous. Kitov's case raises the point that so long as the military did not have to associate its resources with nonmilitary projects, it was content to manage its own internal affairs however it wished. So long as the Party agreed (and often when it did not), it behaved as a private household unto itself. This unpredictability could swing for or against military personnel. In a well-ordered, top-down hierarchical military, research scientists are not usually expected to be able to send letters directly to the heads of state or to influence state policy with those letters. At the same time, a well-ordered military probably would not permit middle-level administrators to dismiss a star scientist from the army for proposing cost-saving procedures that already were supported by the heads of the state, and if such a trial did take place, the appointed dignitaries surely would attend and dismiss the case. Yet none of this happened to Kitov, among untold others—and no one found these events unusual.[28]

The Historical Concurrence of Cold War Networks

Because international communication networks precede national computer networks, multiple network projects often emerge in very different places at about the same time, and priorities are often the last thing to be prioritized. By my accounting, Kitov in the fall of 1959 was the first to propose a national computer network for civilian communication anywhere, although this or any other "first" claim ignores the complex interdependencies of institutions and individuals that create any major technological project. The rush to make "first" claims usually is seen in histories of technological invention (especially histories written by retired professional technologists) to enhance biographical hagiography and ignore claims made elsewhere. It also can be difficult at the edge of any innovation to distinguish between a slight improvement to an old technology and an altogether new technological invention. Kitov's EASU, like most of the proposals examined here, assembled a network out of preexisting and new telegraphy, telephone, radio, and radar networks. Rather than thinking of them as computer networks, EASU was framed more as telephone networks with computers. Simultaneously, the Soviet military, including computer network designer Nikolai Matiukhin, knew of and sought to imitate the automated air defense radar network that went operational in the United States in 1958, although little about the classified SAGE project or its classified Soviet equivalent filtered into civilian science.[29]

Thoughts about ambitious civilian networks were percolating elsewhere as well. Just months after Kitov's second letter, the American psychologist J.C.R. Licklider's 1960 essay "Man-Computer Symbiosis" featured a vision of the potential social and civilian benefits of computers, although (with one footnoted exception) his essay restricts itself to local human-computer intersections. In that footnote, he "envision[s], for a time 10 or 15 years hence, a 'thinking center' that will incorporate the functions of present-day libraries." From here, "the picture readily enlarges itself into a network of such centers, connected to one another by wide-band communication lines and to individual users by leased-wire services. In such a system," Licklider concludes, "the speed of the computers would be balanced, and the cost of the gigantic memories and the sophisticated programs would be divided by the number of users."[30] In 1963, Licklider scaled up his vision of that network as a library with an internal memo that was titled (half in jest) "Memorandum for Members and Affiliates of the Intergalactic Computer Network" and that sketched out the system that became the ARPANET—the technical predecessor to the Internet.

Despite their historical concurrence, all available evidence signposts that the early Soviet economic networks and the ARPANET developed independently of one another. When the ARPANET went online in 1969, it took the Soviet state by surprise. I have encountered no evidence to imply that Kitov or others knew about Western computer network developments other than the SAGE project. Nor have I found evidence that the American secret intelligence community knew about Soviet cybernetic developments before 1964, when Soviet specialists at the CIA began wringing their hands about Soviet cyberneticists working on a nonmilitary "unified information net."[31] Although subsequent Soviets would first pioneer socially ambitious nationwide network projects, the front of Soviet network projects, like the science of cybernetics that underwrote it, proved anything but unified.

Other forms of international influence did lead to networks elsewhere. In October 1957, for example, Soviet authorities set into motion events that led to the ARPANET. Soviet rocket scientists used a missile to launch the first manmade object into terrestrial orbit—*Sputnik I*, the first artificial satellite. At the height of the cold war space race, *Sputnik* came after a number of worrying developments. In November 1955, the Soviets air-dropped their first thermonuclear nuclear bomb, during a time of tension when many American military strategists believed, probably incorrectly, that the Soviet fleet of long-range bombers could reach American targets. The "bomber gap" crisis in the mid-1950s, which was unfounded but drove defense spending, launched that gap into orbital space. With *Sputnik* in orbit, the natural next step was as obvious as it was terrifying: if a warhead were placed atop such satellites, the world could be destroyed in a matter of minutes.

In February 1958, five months after the *Sputnik* crisis, the United States Defense Department created the Advanced Research Projects Agency (ARPA). This new government agency was charged with investing in and advancing the frontiers of technology research beyond the immediate needs of the military, especially in the spheres of space, ballistic missile defense, and nuclear test detection. ARPA did not stay focused on militarizing space for long, however. Two years after its creation, ARPA ceded its space research jurisdiction to the distinctly civilian mission of the National Aeronautics and Space Agency (NASA), which also was founded in 1958. ARPA research then turned toward supporting basic, high-risk, and longterm military research in information processing and computer systems for tracking nuclear threats in the age of *Sputnik*.[32]

The focus on basic computer research questions made ARPA an optimal site—under the auspices of the U.S. Department of Defense, a paragon example of a command-and-control hierarchy—for open-ended basic research. Early computer innovations advanced by ARPA researchers include distributed networking, time sharing, and packet-switching technologies (noted below). In 1965, shortly after President Lyndon B. Johnson called for "creative centers of excellence" to advance basic research among universities, the Department of Defense recommended using the ARPANET to connect preexisting, government-supported computer research sites across the American academy—first at the University of California at Los Angeles, Stanford University, UC Santa Barbara, and the University of Utah and then eastward to the Massachusetts Institute of Technology, Carnegie Mellon University, Harvard University, and other universities.[33] The Soviet military-civilian divide barred similar wide-scale collaboration between defense projects and university contractors.

The ARPANET went online on October 29, 1969, as the first large-scale, dual military-civilian use, packet-switching computer network in the world, the "Mother of all nets" as it has since been known. In its first stage, the ARPANET consisted of leased telephone lines and modems connecting computer terminals at UCLA, Stanford, UC Santa Barbara, and the University of Utah. The first message sent was the prophetic utterance L and O—"lo," not as in "lo and behold" but as in the first two letters of the word *login* that could be sent before the network crashed.[34] ARPA directors in the 1960s negotiated careful balances between Congress (to whom the directors promised research that could be applied to national security issues) and academic research contractors (to whom the directors promised the freedom of basic research that would be independent of any defense rationale).[35]

The heyday of military research in the 1960s came to an end in the political wake of the Vietnam War when in 1969 the first Mansfield amendment curtailed military spending on science across the board and in 1973 the second Mansfield amendment dramatically limited ARPA funding to appropriations for research directly related to military applications. ARPA, stripped of the capacity to do basic research, saw many researchers migrate to a fledgling computer industry, most famously Xerox PARC. Such a brain drain or labor migration from the military to the civilian sector would have had to be directed by military and state oversight in the Soviet Union.

So although in both superpowers the early computing industries in the 1950s through early 1970s depended on military state projects (with private contractors used as spinoffs as well as employed by the U.S. Air Force as its research consultancy), the biggest advantage that the United States wielded over the USSR appears to have less to do with the market independence of the private commerce than the porousness of research, resources, and knowledge flows between military and civilian projects. The modest and mixed military-civilian origins of the ARPANET are worth bearing in mind as well: the ARPANET was designed and launched explicitly for civilian scientists to exchange data at a distance. Its affordances as a network for national communication became obvious after the fact with the invention of email in 1971. At the same time, these civilian, public networked computing utility services were initially funded because of the military justifications to design, fund, and build a nationwide communication network that could survive a nuclear attack by the Soviets. This military motivation led Paul Baran's innovations at RAND in distributed networking and packet-switched networking and distinguished the ARPANET from other networks of its time. The emerging thesis here appears to be that the virtue of the military-industrial-academic complex in the United States rested on not the state, the market, or civilian research but on the *complex* that connected these sectors.[36]

Meanwhile, Chile under Salvador Allende (1970–1973) and France in the 1980s developed large-scale national networks. Unlike the strict Soviet divide between military and civilian research and more like the far more synthesized "military-industrial-academic complex" in America (perhaps the most important element of that phrase for understanding midcentury big science are the hyphens), the cases of Chile and France show that the international history of civilian networks cannot be easily separated from that of military networks. The Soviet military tightly siloed its technical innovations, the East German Stasi shuttered its large-scale computer network capacities from serving and transferring to civilian applications, and the West German government also forbade the transfer of network capacities from military to civilian.[37]

No country escaped institutional frustrations in developing nationwide computer networks. At important times, the complex in cold war American science proved vexatious, if not impossible, to navigate. Take, for example, Paul Baran (1926–2011), a Polish-born engineer who was raised in Philadelphia and Boston. Baran is widely remembered today for innovating packet-switching and distributed-network designs, which now are central to modern-day networking, but his struggles are less well remembered. In 1960 at the RAND Corporation, a research think tank under contract with the U.S. Air Force, Baran articulated the "hot-potato heuristic" behind modern-day data traffic on the Internet: break down a message into packets (or envelopes) of information, release each packet to travel on its own traffic-reducing pathway to its final destination, and resequence and receive all packets in their original order. In the early 1960s, Baran also designed the celebrated idea of a distributed network in which every node in a network connects to its neighboring nodes and not to any decentralized or centralized node arrangement (figure 3.2).

Widely celebrated as a prototype to "end-to-end" intelligence and a liberal democratic mode of communication, Baran's network innovations were colored and shaped by the cold war military complex as well as cybernetic sources. In the embarrassing aftermath of *Sputnik*, the U.S. Defense

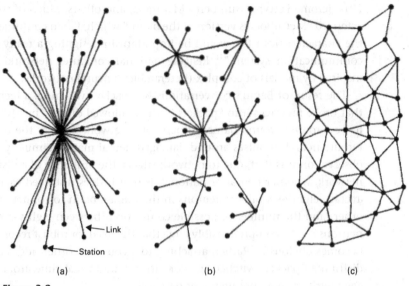

Figure 3.2
Three network types: (a) Centralized, (b) decentralized, and (c) distributed. *Source:* From Paul Baran, "Introduction to Distributed Communication Networks." *On Distributed Communications,* RAND Corporation Memorandum RM-3420-PR, August 1964, 2. Reproduced with permission of The Rand Corp.

Department ordered ARPA to design a "survivable" network that would last long enough in a nuclear strike to send a "go-code" to guarantee "second-strike capability." "There was a clear but not formally stated understanding," noted Baran, "that a survivable communications network is needed to stop, as well as to help avoid, a war."[38] A network that can survive an enemy attack could ensure the threat of the mutual nuclear annihilation—a threat so cataclysmic that it would rationally deter (Baran and his military superiors hoped) either the Soviets, the Americans, or any other nuclear power from striking first.[39]

Baran's inspiration for packet switching as a way to build a survivable network traces back to Warren McCulloch's cybernetic conception of the human brain as a complex and resilient logical processor. As Baran reported in an interview with Stewart Brand, "McCulloch in particular inspired me. He described how he could excise a part of the brain, and the function in that part would move over to another part."[40] The same interview lists McCulloch and Pitt's 1943 paper on neural networks as a sensible reference, although Baran also noted that he was reading more broadly in the "subject of neural nets," a literature that probably included McCulloch,

Pitts, Jerome Lettvin, Humberto Maturana, and others. Much of this characterized "McCulloch's version of the brain," which, Baran continued, "had the characteristics I felt would be important in designing a really reliable communication system."[41] Reliable national computer networking were inspired by models of complex (heterarchical) neural networks.

The result of Baran's conversations was packet switching, a technology that broke messages into "packets," which allowed digital "bursts" of data to be rerouted around damaged parts of a network—just as the brain can reroute neural impulses around damaged neural matter. Similarly, Baran's observation was that, due to network effects, the brilliance of a distributed network, whether neural or national, is that it does not need each of the average eighty-six billion neurons in the human brain to connect to every other (and the number of possible connections between eighty-six billion neurons is so incomprehensibly large that the need for robust reconnection becomes obvious).[42] Rather, attaching to a couple of other nodes allows a distributed packet-switched network to reroute in real time around damaged territory, whether neural or national.

The governing logic behind Baran's innovations is curiously the same as McCulloch's heterarchy: in a heterarchy, the relations between nodes can be ordered and evaluated in more ways than one, and there is no overarching governing structure, no internal logic, and no accounting regime for determining how nodes interconnect. Both lack a fixed control center or mother node. Baran did not concern himself with theorizing about a distributed communication network as a neural network for the nation, as a cyberneticist might. McCulloch's ideas about the brain as a self-governing network helped Baran to arrive at concrete pragmatic solutions to the overarching military orders of his employer. The Internet, in this sense at least, traces its intellectual sources back to cold war cybernetics.

Baran's network innovations do not arrive without serious institutional and international complication. Although technically on target, Baran's ideas were not influential until after a foreigner—an Englishman named Donald Davies, with the UK Post backing him—independently discovered and articulated packet switching. Only then did Baran's superiors in the U.S. military-industrial complex start paying attention to his ideas. In fact, between 1960 and 1966, AT&T repeatedly declined or delayed his proposals to develop digital communication networks. As one AT&T official told him, the near nationwide monopoly on analog telephony networks was not about to go into competition with itself. When it appeared that the air force stood ready to implement Baran's ideas without AT&T, Baran withdrew his proposal because he felt that the appointed government agency, the Defense Communication Agency, would "screw it up and then no one else would be allowed to try, given the failed attempt on the books." With no such "competent organization" in sight and after spending six years aggressively publishing his network research internationally to ensure maximum circulation about how survivable communication networks could help ensure mutual deterrence, Baran despaired at the local prospects and turned his attention elsewhere.[43] The popularity of the phrase *packet switching*, which was Davies's term, and the obscurity of Baran's initial coinage *block switching* are evidence that it took outside competition to spur local authorities to take packet switching seriously. The U.S. ARPANET, despite the efforts of its own network entrepreneurs, was inspired by foreign founders. To the degree that Stigler's law of eponymy holds—"no scientific discovery is named after its original discoverer" (a law that Stigler attributes with a grin to Robert Merton)—Baran's case rehearses not the exception but the rule that international communication networks precede national computer networks.

Aleksandr Kharkevich's Unified Communication System (ESS)

At the same time that Paul Baran was publishing his network research in the hopes of ensuring the Soviets would have access to survivable communication networks and that J.C.R. Licklider was thinking about computer networks as pragmatic tools for facilitating long-distance exchange of scientific data, Soviet cybernetic network entrepreneurs were imagining computer networks as ambitious infrastructural solutions to the nation's most pressing civilian problems. In the imaginative minds of the three Soviet cyberneticists chronicled below, digital computer networks were models both of and for the entire nation.[44] These and other early proposals for a "unified system of calculating centers for the development of economic information" found their earliest inspiration in the 1955 Academy of Sciences proposal by Vasily Nemchinov (two years before *Sputnik* and well before the invention of the ARPANET) that considered erecting large but unconnected state computer centers (in Moscow, Kiev, Novosibirsk, Riga, Kharkov, and other major cities) that could facilitate the local exchange of scientific reports and economic information among regional economists. One of those proposals—Kharkevich's unified all-state system for information transmission—has been relatively neglected in previous commentary and receives additional attention below.

In 1962, Aleksandr A. Kharkevich, then deputy chair of the Council on Cybernetics, proposed a communications network for the entire nation, although this proposal—for a network that was formally called the "unified all-government system for the transmission of information" (*edinaya obsh-chegosudarstvennaya sistema peredachi informatsii*)—did not seek to solve an explicit civilian-sector problem, unlike other contemporary Soviet network projects. In fact it sought to solve no particular problem at all: it was proposed out of sheer technical ambition to build a national communication network on preexisting telephony and telegraphic channels for all kinds of data exchange. The closest that Kharkevich comes to a social justification is noting, without comment, that it will "broaden the sphere of human activity."[45] The technical orientation of data exchange in Kharkevich's proposal resembles the purpose of the ARPANET as a network for exchanging data between scientists. With similar "intergalactic" ambitions, Kharkevich set out to optimize all technical communication problems at once by proposing to merge all Soviet data streams into a single nationwide digital communication network. His 1962 proposal came to light in an article titled "Information and Technology" that was published in the leading periodical *Communist*, in which Kharkevich apparently renamed this network with the more workable title of "unified communication system" (ESS, for *edinaya sistema svyazi*), a possible source of the uncited CIA speculations about a menacing Soviet "unified information net."[46] His vision describes a technical future that was obvious to information theorists, who were the technocratic twin of cyberneticists and could be traced back to Claude Shannon of Bell Labs and his seminal 1948 article "A Mathematical Theory of Communication." (Kharkevich was himself a leading information theorist and specialist in noise reduction in electronic communication signals.)[47]

In the 1962 *Communist* article, Kharkevich proposes that the ESS unified network of information transmission be built, like the other proposals here, on the preexisting telephone and electronic network infrastructure, which he found analogous to a nationwide railway network that was built to transmit, store, and process digital information messages. Given that most Soviet citizens had to use public phones in 1962, this was a fantastically far-fetched technical proposal on any terms. Perhaps for this reason, he devotes almost the entire twelve-page article to the technical capacities of such a network (for example, how messages would arrive at the right place without data loss), which were grandiose. Telegraph cables, telephone lines, radio waves, and all other technical communication channels were to be unified into a common digital and "enciphered" currency that would be related by binary electronic pulses over telephone wires. The irreducible denominator to his technocratic vision was the concept of information: "the far-reaching role of information has become clear not only in the relations between people, but also in the interactions between man and machine, as well as in the life of any organism." He continued, "with the enhancement of economic, technical, and cultural levels of society, the amount of information necessary to collect, transmit, and somehow provide for all functions of the community of people grows faster and faster. No organized form of activity is thinkable without information exchange. Without information, planning and governance are impossible."

Backlit by the stated universal need for information, Kharkevich justified the network proposal by citing the "prominent system 'SAGE'" computer system in the United States and Canada as a parallel to his vision of a narrowly applied, universal information system for antiaircraft defense. The top of his pyramid, ESS network design, was meant to "fulfill the function of the dispatcher of the network," or "the center will be constituted by a large group of specialized calculating-logic machines, appointed for the direct resolution of the many changing conditions of one single task: the supply of increasingly favorable conditions for the appointment of all currency flows of information."[48] The Soviet Union did need not to stop at antiaircraft defense, he said, concluding his "grandiose thought" of an all-reaching, full-service ESS network with the observation that "creating an all-state unified system of connection ... would only be possible in a socialistic government under the conditions of a planned economy and centralized government."[49]

Kharkevich's article is remembered among some technologists today not for proposing the ESS but for formulating what became known as Kharkevich's law. This law holds that the quantity of information in a country grows proportionally to the square of the industrial potential of the country (N^2). The original formulation of his observation in the article is perhaps less elegant than information technologists might remember: "Given a large number of factories, the number of paired links between them is approximately equal to half of the square of the number of factories."[50] The law, in effect, prophesies a power law connection at the macro level between an industrial society and an information society. In 1965, the American computer businessman Gordon Moore expressed a distinct exponential law that has applied to the microscopic level of the compounding growth of silicon chip production—that the number of transistors on an integrated circuit doubles every two years (2^N).[51] Both men foresaw in 1962 the emerging information sector or what Austrian American economist Fritz Machlup called "the knowledge economy." For Kharkevich, the amount of information that a society processes can be expressed as a power law function of the industries it contains, and for Moore, the amount of information that a society processes can be expressed as an exponential function of the transistors on the circuits its industries can produce.[52] These sibling laws (Moore's 2^N and Kharkevich's N^2) diverge interestingly in complex systems (when N is larger than 4). They also backlight their micro and macro focuses—Moore on microscopic industrial production and Kharkevich on informational industrial society. The result is different framings of the national network as a sort of central processor. Like Baran, Kharkevich prioritizes building "survivable" military networks while also looking to benefit other civilian and social goals.

Unlike the cybernetic metaphor of the brain that Baran drew from Warren McCulloch, Kharkevich saw his national network as a nervous system that was overlaid onto the body of the nation and that would be governed by a central processor, or brain, located in Moscow. The resulting contrast of cybernetic metaphors for the information societies is again sharp: for Kharkevich, the networked nation was the body controlled by a central brain, and for Baran, the networked nation was the brain itself.

Like other network designers, Kharkevich also designed the ESS network after the formal administrative structure of the nation that he imagined it would network. "It is natural that the network should be supervised," he wrote, "by the Ministry of Communication [Svyaz'] in the Soviet Union," the ministry that managed many preexisting networks for information exchange, including telegraph, telephone, phototelegraph, messages (courier), and early digital technologies then available in small numbers. Kharkevich breezily dismissed the distributed network model that Paul Baran was developing at the time (although not by that name), observing that the structure of the network needs to be able to connect any two nodes, and he writes, "in order to do this, it goes without saying, one does not need to unite all nodes with separate lines."[53] Instead, Kharkevich considered a hierarchically decentralized design, or pyramid structure, "the rational structure of a network."[54] Like transport roads, his network would split out in a "radial system" in which a "given territorial group is united by links to a communication node." Just as every local, regional, and territorial group would have its own common node, Kharkevich was quick to stress the center that was implicit in this "radial" design. In 1962 in Moscow, a "centralized automated management" design would have appeared reasonable, if still monumental in aspiration, to him:

The brisk carrying out of these functions is possible only ... if the entire network will work under centralized automated management. The governing center of ESS should distribute information about the state of the network at every given moment.... The center should be capable of predicting such changes [in the network traffic] slightly in advance and create the needed operating reserves. The center appoints the pathways for the passage of flows of information as they depend on the general state of the network; in the case of necessity [*nadobnost'*], the center will be able to focus all network resources on fulfilling special information transmission tasks. In short, the governing center of ESS fulfills the function of the dispatching manager of the network.[55]

Here, without the benefit of packet-switching protocols, Kharkevich anticipated the needs of a nationwide network to adapt automatically in real time to traffic jams as well as the capacity to complete "special information transmission tasks" (the sending of nuclear "go-codes" in the case of nuclear "necessity"). Automation appears to be the ultimate nuclear safeguard, for he continues that "it will not be possible to give these functions to people. The center will constitute a large group of specialized calculating-logical machines, appointed for the direct resolution of the changing conditions of a single task: providing increasingly favorable conditions for the appointment of current flows of information."[56]

The fate of the ESS owes less to the technicalities of its design than to the muses of institutional historical contingency. In 1961, the year before he proposed the ESS, Kharkevich was made director of the Institute for the Problems of the Transmission of Information (IPPI), the new Soviet Academy of Science's research center on information technology. The president of the Academy of Sciences, Mstislav Keldysh, a rocket scientist and mathematician who helped develop the Calculation Bureau during World War II, created Kharkevich's IPPI in the same year that he created Glushkov's Institute of Cybernetics in Kiev and Fedorenko's Central Economic-Mathematical Institute in Moscow. In 1963, Kharkevich's ESS vision took its first step forward when the Ministry of Communication created an interagency Coordinating Council, chaired by the then minister of communications, General-Colonel N. D. Psurtseva, to supervise the creation and standardization protocols for the ESS. However, before any the council could make concrete progress and three years after proposing the ESS, the project collapsed. On March 30, 1965, Kharkevich died of protracted health problems at the age of sixty-one. Why no one took up his reigns on the ESS proposal remains unclear, although the lack of evidence implies that the ESS's political prospects passed into history with Kharkevich.

N. I. Kovalev's Rational System for Economic Control

Consider still another short-lived and concurrent network proposal, whose fate archival materials and interviews have not yet clarified. In the

November 1962 Plenary Meeting of the Communist Central Committee, decisions were made to mechanize and automate both the industrial processes and the administrative control over those processes. In the 1963 issue of *Problems of Economic Transition*, N. I. Kovalev, then the director of the State Economic Council (*Goseconomsovet*), published a proposal that elaborated on those decisions and proposed creating and connecting the preexisting major computing centers for each of the regional economic councils (*sovnarkhozy*) that Khrushchev initiated in 1957. Like all the others, Kovalev's design also mapped a pyramid communication network onto the economy's three-tier hierarchy of ministry, regional council, and local enterprise. The network was meant to help the regional councils to receive otherwise unspecified "necessary information" on time. No longer would "the report materials arrive so late that they cannot be effectively used to plan and govern the national economy."[57] Citing Nemchinov and Glushkov (prominent specialists in the field who are featured in the next chapter), Kovalev estimated that the network would cost 94 million rubles, the first layer of thirty computing center would require three years to complete, and the economic savings would far outweighing the costs.[58] By referring to such a computer network as a "rational system," Kovalev did not emphasize the transformative effects of long-distance real-time computer networks but instead stated a need for vaguely specified "cybernetics, electronic computing and control devices" to serve as the "material and technical base" for a transition to a communist model for "planning and controlling the economy" over the next two decades.

Kovalev's proposal stands as a synecdoche for a larger competition among the cybernetic and mathematical economists on one side and state planning agencies and party leaders on the other. Both economic planners and party leaders advanced arguments for and against the computerization and networking of the command economy in terms of whether technocratic reform would lead to the proper control over information. Kovalev, together with his cybernetic colleagues and allies, saw in the networked computer a grand manipulator for transforming the economy as a giant information system in need of optimization, objective planning, and diminishing bureaucratic overhead costs. Curiously, the most influential opposition to such proposals came from the main planning state agencies, including Gosplan, the Central Statistical Administration (CSA), Gossnab, and regional and branch committees. These groups openly resisted his and similar proposals because they were perceived to involve personal loss of control over the information in the command economy.

Conclusion

By 1963, with three national network proposals already on the table—Kitov's EASU, Kharkevich's ESS, and Kovalev's unabbreviated "rational system of economic control"—the institutional landscape was evolving toward some kind of head in economic reform. That intellectual terrain includes various supplemental network projects that promoted the core Soviet cybernetic instinct that large-scale information systems, such as the command economy, can become self-sustaining and even self-governing systems. It may be helpful to distinguish between two meanings of the word *automated*—(1) having operations that are entirely independent of human involvement and (2) having operations that are designed to receive and interact with humans but do not necessarily need human involvement. The OGAS, understood as an explicitly cybernetic human-computer interface, clearly signals the latter sense of the term.[59] In other words, the conceit of cybernetic (human-machine) self-sufficiency was not to imagine a national economy that was independent of any other outside forces but rather to envision a socialist planning apparatus that engaged with the economic body it networked and that, together, would prove responsive, balanced, and self-governing. By contrast, the liberal economists sought a different path to self-governing markets—introducing profit measures into local enterprise accounting while still maintaining basic production guidelines for the overall economy. Both cybernetic and economic liberal reforms reached compromise solutions with the operations of the command economy, just in opposite directions. The cybernetic economists offered a technocratic reform that was meant to work with human administrators and liberal economists—a market reform that was meant to work with command economy guidelines.

These contending approaches came to a head in 1963 through 1965 at the same time as the bumpy transition of state power from Khrushchev to Brezhnev. Because both approaches to reform met with unsystematic but widespread resistance from orthodox economic planners and professionals who were comfortable in their current positions, both produced tentative heirs to the economic debates in the early-mid 1960s—the OGAS proposal in 1963 and the Kosygin-Liberman reforms of 1965. Early Soviet networked computing culture was decentralized in practice, despite the state's centralized design in principle. Kitov's EASU first proposed having technomilitary networks be put to public and social benefit, but he found himself grounded for attempting to bridge the yawning military-civilian divide. Three years later, Kharkevich's ESS, with Kovalev following suit, reached ever further, networking all technical signals into a network resembling the pyramidal state while staying silent about any social ambitions. Yet these early proposals fell prey to strategic veto points in the state administration that depended not on bureaucratic rules but on charismatic leadership and personal power. The technical open-endedness of Kharkevich's ESS probably most closely resembles that of the ARPANET, although the ARPANET began with the modest goal of scientific data exchange and the ESS, like the others, began with an ambitious blueprint for an entire digital nation. Unsurprisingly, the more ideologically charged economic networks faced more ideological opposition, and the fate of the ESS points to the charismatic actor-dependent institutional disorder that governed the Soviet knowledge base.

Perhaps the signal lesson to take from these early Soviet network proposals is that there is no inherent connection between the designs of technological and political systems. Many digital theorists in liberal democracies have imagined the effects of technology in the terms of their local political systems, claiming that digital technologies must be deliberative, direct, and participatory—similar to that of contemporary democracy discourse. So, too, did these Soviet cybernetic theorists imagine that a nationwide computer network would "naturally" map onto the design biases and design logics of the formally top-down centralized administrative hierarchy of the Soviet state. Both visions are theoretically imaginative because they neglect actual political practices and their significant costs and consequences. These network proposals ignored the informal, nonhierarchical functions of the Soviet state and society, just as modern democracies involve far more than just the representation of individual voices celebrated by many digital media theorists. Centralizing computer networks and centralized socialist states have as little to do with one another as the digital does with democracy. Both propose imaginatively rich associations about what could be, promising no less than some pseudo-automatic or pseudo-democratic form of self-determination, but do little to affect careful or accurate assessments of how politics actually works on the ground.[60]

These waves of cybernetic imagination about the fit between computer network and formal state and social structure repeatedly broke against the rocks of widespread practice that countered the official Soviet imagination of itself. Paul Baran struggled to secure institutional support from American corporations and a state that refused to recognize the value of what became the key network innovations of his age. So, too, in Moscow, Kitov had to abort the EASU due to his unsuccessful attempt to bridge the abyss that separated military and civilian research, Kharkevich's ESS collapsed with the health of the man appointed to steer its grandiose technical ambitions, and Kovalev's "rational system" fell short of convincing his colleagues in economic administration that ceding their own decision-making power to automated computers could either rationalize or systematize the work of economic planning. All four of these early network projects—three in the Soviet Union and one in the United States—did not take shape due to an imagined and often misleading connection between political and technical systems. All four rooted their imaginations in the explicitly cybernetic terms of analogizing across technological and social systems. This imaginative and at times utopian instinct for political-technological system analogs leads theorists to neglect the significant costs and consequences that come from actual political practice. As it often happens, the revolutionary reach of our modern technological imagination of large-scale networks (among other things) often ends up serving local institutional self-interests and the status quo. The next chapter extends and complicates this theme in its history and analysis of the central and longest-lasting attempt to network the Soviet Union.

4 Staging the OGAS, 1962 to 1969

The year 1962 proved to be a tumultuous one for the world. Khrushchev's grasp on the reigns of the Soviet state began to slip in the face of mounting criticism, and Kennedy's Bay of Pigs invasion metastasized into the Cuban missile crisis, probably the closest the world has yet come to a nuclear world war.[1] Behind the scenes to these potentially cataclysmic situations, a small team of Soviet cyberneticists who were located in Kiev and Moscow were committed to building "electronic socialism" under the guise of the All-State Automated System, or OGAS. The OGAS Project was the Soviet Union's attempt to build a national computer network project that would network the command economy, automate and optimize the immense coordination problems besetting that economy, and thereby speed the grand socialist experiment toward a prosperous and stable Communist future.

The All-State Automated System Project took its first breath with the delivery of a sealed envelope into the hand of Nikita Khrushchev in the late fall of 1962. The letter to the general secretary was written by young scientists from the Komsomol Spotlight (*Komsomol'skii prozhektor*), who noted what they perceived to be the catastrophic backwardness of information technology in the USSR compared to the United States and called for the immediate acceleration and adoption of computing technology into economic planning. The letter made an impression on the public, in the form of an official *Izvestiya* newspaper article titled "Information Technology in the National Economy," and on members of the Politburo, the governing committee of the Soviet state, which reportedly spent nearly thirty-five minutes of a forty-five-minute session discussing the consequences of their fifteen-page letter. Several months later, on May 21, 1963, following the proposals discussed in the previous chapter, the Politburo with the backing of all relevant ministers advanced a Communist Party resolution calling for the same and authorizing the first economic reform carried out by automated computer network (later known as the OGAS). This chapter discusses the vision, the chief visionary (Viktor Glushkov) and his team, and the institutional landscape for the OGAS Project, the most prominent attempt to establish a civilian national network project in the Soviet Union.

The OGAS: A Vast Vision behind a Global-Local Network

The OGAS Project promised to deliver "electronic socialism" that was as ambitious as its official title was long—the All-State Automated System for the Gathering and Processing of Information for the Accounting, Planning, and Governance of the National Economy, USSR. Its short names were the All-State Automated System for the Management of the Economy, the All-State Automated System, and OGAS. For clarity, I distinguish here between the OGAS as the imagined network that did not come to exist and the OGAS Project as the Soviet actors and institutions that tried to realize this reform.

According to its cyberneticist founders, the infrastructure of the command economy had to be upgraded before the entrenched coordination problems that led to the country's economic woes could be resolved. "In the area of economic management," Glushkov wrote in 1962, "cybernetics fits our socialist planned economy like a glove." The work was fundamentally technocratic and rational and sought to "reduce the influence of the subjective factor in the making of administrative decisions."[2]

In its most modest framing, the OGAS—which stretched nationwide across preexisting and new telephony wires that were entirely separate from preexisting military computer networks—appears little more than the extension of a local factory control computer network. It would be an ASU (automated system of management) or OGASU (All-State ASU) (*Obshche-Gosudarstvennaya Avtomatizirovannya Sistema Upravleniya*). The primary visionary of the OGAS, Viktor Glushkov (who is discussed later in this chapter), had been aware of Anatoly Kitov's efforts, including his Red Book letter, ever since Glushkov began studying computing in Kiev with Kitov's 1956 *Digital Computing Machines* in hand. Glushkov employed Kitov as a consultant in 1960 after Kitov's dismissal from the army. The OGAS was to become the Soviet equivalent of the national economy imagined as a single factory, with one interactive industrial control system serving it across a national computer network in real time. This was not to be a dumb network that would merely exchange data and communication across great distances. It was to be a "smart" network whose decentralized command and control protocols would be capable of automating, mathematically modeling, optimizing, and rationalizing away the profound inefficiencies that beset the command economy. According to the original proposers, the resulting network efficiencies, which were optimized to serve both national and local needs, would achieve full effect by 1990, nearly thirty years after the Komsomol young scientists delivered their letter into Khrushchev's hands.

As originally envisioned, the OGAS had several distinct features. Perhaps the most meaningful contrast with that of modern networks is that the OGAS was modeled after the economy of a factory writ large for a nation. The basic unit of the OGASU was, as the initials imply, the ASU, the automated management system, or a local information and control system that looped onsite mainframe computers into the industrial processes of a factory or enterprise to provide real-time information feedback, control, and efficiencies. This kernel vision of a network as an expression of the nervous system of a factory, writ large across a nation, magnified the image of the workplace until it incorporated the whole command economy—a sort of simultaneously metaphorical and mechanical collectivization of the industrial household (or what Hannah Arendt calls the *oikos*).

The OGAS Project might be seen as preceding, although not precipitating, the current trends in so-called cloud computing. The national network was to provide "collective access," "remote access," and "distance access" on a massive scale to civilian users who could "access," "input," "receive," and "process" data related to the command economy (such older terms appear to bear more descriptive heft than the modern computing metaphors such as *upload, download, share,* and *stream*). The decentralized network was designed so that information for economic planning could be transmitted, modified, and managed in relative real time up, down, and laterally across the networked administrative pyramid. At the base of that pyramid, in the network's initial vision, were as many as twenty thousand computer access points and ASUs distributed throughout the nation's enterprises and factories. This base of computer centers would be connected to one hundred to two hundred midlevel regional planning decision centers in major cities, which would be connected to the central planning processing center in Moscow by high-capacity data channels. The original vision of a three-tiered pyramid network—with twenty thousand computer centers on the bottom, one hundred to two hundred in the middle, and one on the top—was scaled back in the original design of the technical base of that network (the Unified State Network of Computing Centers, or EGSVT). The first proposal for that technical network offered a modest blueprint where one central computing center in Moscow would regulate only twenty-five to thirty computing centers in city sites of "information flow concentrations" and an unspecified number of "regional calculating center and points of information gathering"[3] (figures 4.1 and 4.2).

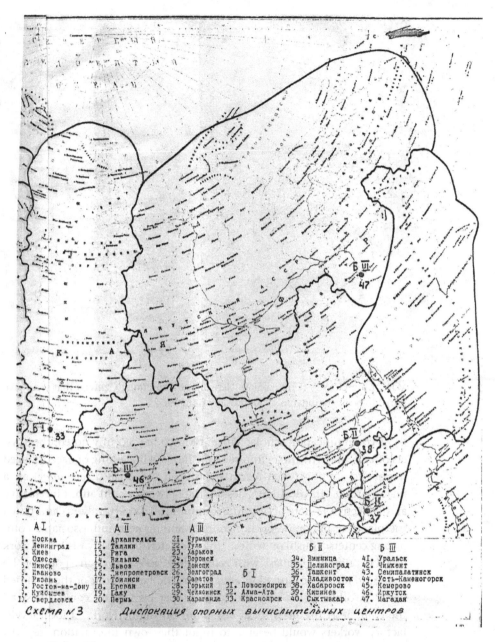

Figure 4.1
Map of the three tiers (I, II, III) of planned computing center sites behind the OGAS (All-State Automated System), 1964.

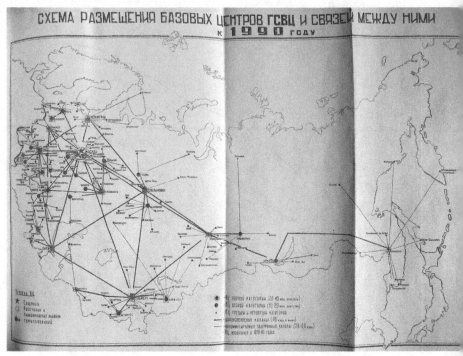

Figure 4.2
Map of the EGSVTs (Unified State Network of Computing Centers) that were projected to be operational in 1990, possibly from 1964.[4]

As communication scholar Vincent Mosco has recently noted, the Soviets offer perhaps the first glimpse of the modern imagining of decentralized remote computing (what recently has been called *cloud computing*) on a massive scale.[5] In Glushkov's design, the network would afford interactive and collective remote access and communication vertically up and down the planning pyramid and horizontally among peer and associated computing centers. Glushkov writes: "the characteristic quality of the network was a distributed database with zero-address access from any point of the system to all the information after automatic verification of the qualified user." In other words, any user with proper permission could access all the content of the network at any point on the network. At local levels, factory workers would be able to input their own information, reports, and recommendations about improving factory workflow, which would automatically be stored in a national unified database for local, regional, and national review. The content format was not to be prespecified. For example, the network visionaries planned to include over 500,000 project dossiers on foreign scientists, engineers, executives, and companies in the OGAS nationally networked database. (From 1963 to 1968, the associated Department of Scientific Institutions gathered about 75,000 such dossiers.) The proposal's other ambitions went far beyond that of simply sluicing economic planning information. In 1971, the deputy editor of *Pravda*, Viktor Afanasyev, for example, reasoned that OGAS "can be used—and should be used—for gathering, processing, and analyzing information on sociopolitical and ideological processes as well, for the purpose of optimal management [of society]."[6]

As a near synthesis of optimal management and total surveillance, the OGAS is a full articulation of the wider political-economic imagination of the Soviet Union as not just a single unified society and set of nations but as a unified corporation with a socialist mission statement. The OGAS appeared to its founders as the information technology upgrade that the Soviet Union had long needed to be able to function as the corporation it had already long imagined its command economy to be—a single and complex organization that featured decentralized means of control and communication for circulating the informatics lifeblood of a socialist economy. Because socialism openly recognized economic activity as more than merely computational, the network that would best facilitate its fitness would also control and communicate associated political and social concerns as well.

The OGAS Project of course was no ARPANET. It sought much more than data transfer and communication among scientists. From the outset, the OGAS Project sought to bring the economic bureaucracy online by making all relevant government documents electronic, allowing a decentralized remote access to all economic workers, and allowing decentralized access for controlling and optimizing the information in those documents. The decentralized design of the network project is worth stressing. Although still hierarchical, acquiescent to Moscow as the center, and state-led, the longest-lasting Soviet network proposal was (unlike the full central control in Kitov's EASU and the radial design of Kharkevich's ESS) openly worker-oriented, antibureaucratic, and decentralizing in principle. This gives the OGAS Project and its team more credit than many commentators and critics have given it. Both international and internal critics, including the British organizational cyberneticist Stafford Beer, were critical of Soviet management techniques.[7] More than a network, the OGAS Project as formulated by Glushkov outlines a daring technocratic economic imagining that was meant to operate in a future Soviet information society by digitizing, supervising, and optimizing the coordination challenges besetting the national command economy.

The associated costs and scale of such a supercharged system were accordingly colossal. Glushkov captured the sentiment of network effects, which is still alive in surveillance capitalism's promotion of big data today, in this phrase: "world practice shows that the larger the object for which an information-management system is created, the greater its economic effect."[8] More than *komchamstvo*, or Lenin's term for "Communist boasting," the basic OGAS blueprint affirms its staggering magnitude. In its initial proposals, the OGAS Project estimated that it would take over thirty years to be fully online, that it would need a labor transfer of some 300,000 personnel, that costs would be upward of 20 billion rubles for the first fifteen years, and that tens of thousands of computing center and interactive access points would be distributed across the Soviet population.

All this would prove net efficient, promised Glushkov. The 300,000 knowledge workers would constitute an enormous labor transfer, as well as a net reduction in the ever-rising number of people who were employed in economic planning. The 20 billion rubles would be distributed over three five-year plans, with the first requiring a seemingly modest 5 billion rubles. Acutely aware of the advantages of the well-regulated financial management that was enjoyed by the successful military nuclear and space programs, Glushkov insisted to Prime Minister Kosygin that, if the OGAS were to be developed, this civilian program would require a similarly well-managed funding stream, even though it would prove more complicated and expensive than both military programs combined. For his distinctly decentralized civilian economic communication infrastructure project, Glushkov sought fully centralized military-style financial funding. Only with well-managed funding could this civilian project pay for itself, which it promised to do handsomely, returning fivefold on the first fifteen-year investment, or "no less than 100 billion rubles" (roughly $850 billion in 2016 U.S. dollars), and even this estimated windfall in savings "was a conservative figure."

Cost, in other words, is the simplest reason that the OGAS Project never developed as proposed. A networked command economy, as economist critics noted, would simply prove uneconomical. No such sum of funding was granted, and the projected costs soared slowly upward until, according to varying estimates, the OGAS, if built in the late Soviet Union, would cost the staggering sum of 160 billion rubles (or $1.4 trillion in 2016 dollars, or roughly the U.S. deficit in 2009).[9] Still, costs are never black or white. The OGAS Project imagined a series of adjunct projects with less painful price tags. As early as 1963, the EGSVT technical network proposed a far more affordable fraction of this vision—one center in Moscow, twenty to thirty regional computing centers, and unspecified local computing "access points."

The Visionary behind the Vast Network: Viktor Glushkov

Viktor Glushkov (1923–1982), who was called the "king of Soviet cybernetics" in his *New York Times* obituary, was neither the first nor the last to propose a nationwide network. But he figures as the organizing protagonist of the remaining history as the leading champion of the OGAS Project, a well-positioned academician, vice president of the Academy of Sciences, and a leading cyberneticist. Known as both a global thinker and a local doer from a young age, Viktor Mikhailovich Glushkov was born in the temperate southern Russian city of Rostov-on-Don on August 23, 1923, into a family of a mining engineer (figure 4.3). Like many prominent Soviet figures, he excelled in mathematics at a young age and in middle school dreamed of becoming a theoretical physicist. In high school, he quickly grasped topics such as quantum mechanics and absorbed classics in the original German from Johann von Goethe to Georg Wilhelm Friedrich Hegel's *The Philosophy of History*. In 1941, the Nazis executed his mother for her part in the underground resistance. After failing to enlist in the artillery school for health reasons, he turned to mathematics in college, dove into topological algebra, and graduated in 1948. Four years later, including two years to complete his doctorate while holding a research position at a new nuclear center in Yekaterinburg (then Sverdlovsk) in central Russia, he proposed solutions to David Hilbert's generalized fifth problem in 1952. In 1900, in Paris, the German mathematician David Hilbert proposed twenty-three foundational problems that have attracted much attention in modern mathematics since. Two of those problems are considered unresolvable, and the fifth problem, parts of which Glushkov tackled, involves smooth manifolds in Lie group theory. The initial breakthrough came to him while he was climbing an ice field on Mt. Kazbek in the Caucasus with his wife, Valentina Mikhailovna. Six months later he had formalized the shortest solution to that problem to that day.

This feat guaranteed that in the mid-1950s, the rising algebraist could have secured almost any position in the Soviet Union. Thanks to an introduction from academician Boris Vladimirovich Gnedenko, Glushkov became acquainted with Lebedev's computing center in Kiev, which six years later (in 1962) he transformed into the prominent Institute of Cybernetics. He directed the institute from 1962 until his death in 1983. When asked why he chose to shift his attention to the intersection of computer technology and mathematics and subsequently assume the directorship of a Computing Center from Sergei Lebedev in Kiev, Ukraine, in 1956—and not a more politically prestigious position in Moscow—he is reported to have replied that his wife, Valentina, whom he had met in their third year

Figure 4.3
Viktor Glushkov, about 1963. From the personal
archives of Viktor Mikhailovich Glushkov.

of college in relatively balmy Rostov, preferred the warmer weather in Kiev
and he agreed.[10] It is also possible that this committed theorist of decentralized power saw a position removed from Moscow as a strategic opportunity to practice and leverage decentralized power. So having achieved an
ambitious goal in mathematics at a young age in 1956, Glushkov turned
his sights to theorizing the emergent field of cybernetics, especially the
relationships between information technology and economic cybernetics. His oeuvre swept across theoretical fields (including abstract algebra,
mathematical logic, automata theory, and algorithms) and applied fields
(including the development of hardware, software, robotics, informatics,
and computers and the administration of Soviet economic cybernetics).

Glushkov is remembered by colleagues for having been always "on"—a
persistent kind of applied grand theorist—except for the occasional hike or
fishing trip down the Dnieper River, which he relished. His children recall
him following a strict daily regime: when he was not riding the day-long
Kiev-Moscow train (which he jokingly called his home), he rose at 8:30

am, exercised, breakfasted, went to work, returned home in the evening, and continued working until about 2:00 or 3:00 am. In 1963, as the first director of the brand new Institute of Cybernetics, his work habits reached a feverish pitch and then broke. Valentina recalled that he worked eighteen to twenty hours a day until at age forty, he suddenly collapsed from a brain seizure. (A tumor of the medulla likely ended his life twenty years later.) Undeterred and still bound to his hospital bed, he finished the introduction to his Lenin Prize–winning book, *The Design of Digital Automatic Machines*. His intense persistence of mind rendered possible his mathematical achievements, and his vision was shortsighted since youth due to his voracious reading habits. It is not known whether this contributed to his protracted struggle with a fatal brain tumor.

Fluent enough in German and English to lecture and publish abroad in those languages (having once recited excerpts from Goethe from memory for two hours to win a bet), Glushkov figures as a consummate information universalist, even among cyberneticists, for whom practically every challenge reduced to, as his colleague and fellow computer pioneer Boris Malinovsky put it, "the global problem of the computerization of information sharing."[11] Committed to building computer networks that share information, his subsequent research goals pushed him to generalize his applied innovations further and further. Some characteristic examples include career examinations in not just specialized computing but multipurpose control computing; not just von Neumann computing processor architecture but "massively parallel macro-piping" and a recursive base for fractal processing in computer architecture; not just computer programming but natural-language computer programming; not just robots but entirely digital automata; not just bureaucracy but paperless offices and informatics; and in the end, not just a better economic life for private humans and our collective humankind but an even bolder and more remote future. He identified in the inevitable evolution of artificial intelligence the possibility of "informational immortality," where the subjective consciousness, memories, and personalities of individuals and societies might be transferred into a global network that was capable of outlasting the ages, resurrecting and recasting civilization as we know it.[12] Because they reach so far, the endpoints of these various research initiatives begin to express in relief the grander vision that organized his personal commitment to the OGAS Project as the next step in networking onto the higher plane of the grand collective of socialist labor.[13] For Glushkov, the OGAS Project represented a vehicle for achieving the whole of his many scalable visions.

Glushkov modeled his thinking about computer networks and processing after—and often against—the prevailing trends in the study of neural networks. In a notable deviation from von Neumann digital computer architecture that pushes all data bits through a bottleneck one bit at a time, Glushkov theorized about what he called the "macropiping" or "macroconveyor" processor architecture for transmitting information along multiple processors simultaneously between groups of computers. Macropiping was modeled after his cybernetic vision of the computer, which, according to a 1959 speech, would best resemble the human brain in its capacity to process billions of bits of data in parallel simultaneity. This idea germinated into his notion of a simultaneous national network that would function as a self-regulating nervous system for the whole of the Soviet people. Glushkov shared conversations and computing technology with people such as chessmaster Mikhail Botvinnik, among many other ambitious dreamers, to create a machine in the image of man, not the other way around. In the late 1950s, Glushkov sought to develop in theory a computer programming that imitated the sophistication of human thought, cognitive function, and natural language. For example, he and his colleagues examined processes for distinguishing between grammatically and semantically correct sentences, such as "The chair stood on the ceiling," as a step toward achieving natural language programming and a more human "higher intellect" in the computer.[14]

The OGAS Project took shape in a complex network of research teams (at the center of which sat Glushkov). No science is a solitary endeavor, however, and a full accounting of the details of the people who constituted Glushkov's teams, their accomplishments, and their frustrations is beyond the scope of this book. Two of his favorite students and eventually a wife-husband team, Yulia Kapitonova and Aleksandr Letichevsky, identify what they call the intellectual "school" of Viktor Glushkov, which itself contained many teams that contributed to the OGAS Project and many other projects. The first EGSVT proposal began to take shape in the conversations of Glushkov, Vladimir S. Mikhalevich (who directed the Institute after Glushkov), Anatoly Kitov, A. Nikitin, and others, and the first government document published on the EGSVT, on May 21, 1963, also highlights as coauthors Anatoly Kitov, V. Purgachev, Yu. Chernyak, M. Popov, among others. Key members and colleagues at the Institute of Cybernetics in Kiev included Vladimir S. Mikhalevich, V. I. Skurikhin, A. A. Morozov, Yulia V. Kapitonova, Aleksandr A. Letichevsky, A. A. Stognii, T. P. Mar'yanovich, and others. The Moscow-based supporting scientists included Anatoly Kitov, Yu. A. Antipov, I. A. Danil'chenko, Yu. A. Mikheev, R. A. Mikheeva, among others.[15] Optimization modeling, which would have contributed to the management software running the OGAS economic reform) were developed from 1962 to 1969 by Vladimir Mikhalevich, O. O. Bakaev, Yu. M. Ermol'ev, I. V. Sergienko, V. L. Volkovich, B. M. Pshenychniyi, V. V. Shkurba, N. Z. Shor, and others. Glushkov toiled alongside A. A. Stognii and A. G. Kukharchuk as principal designer in developing the Dnepr-2, a transistor computer. He also headed a team that included Y. Blagoveshensky, Aleksandr A. Letichevsky, V. Losev, I. Mochanov, S. Pogrebinsky, and A. A. Stognii in developing the MIR-1 engineering calculation machine, an exhibition version of which IBM purchased in London.[16]

Other supporting teams in the Glushkov school indirectly reflect on the OGAS Project. Kapitonova and Letichevsky, for example, helped Glushkov theorize an "analytic" mathematical human language programming language and an algorithmic design in computer design automation.[17] This team helped nudge the field of artificial intelligence away from the notion that the brain was machine-inspired (away from McCulloch and Pitts's claim that the brain follows logical circuitry). Instead, they worked on building a brain-inspired machine that was "capable of carrying out complex creative activities," continuously seeking to reveal the "higher intellect" of machines modeled after mechanisms of the mind.[18] If there was a danger in the brain and machine metaphor, it ran only one way for Glushkov: "the danger is not that machines will begin to think like people," he intoned, "but that people will begin to think like machines."[19]

Are National Networks More Like Brains or Nervous Systems?

In 1962, Glushkov imagined the OGAS as a "brainlike" (*mozgopodnobyi*) network for managing the national economy and extending the life experience of the nation and its inhabitants. Consider the implications for the cybernetic analog between neural networks and national computer networks. As already noted, cybernetics brings to bear powerful conceptual frameworks for imagining structural analogies between ontologically different information systems—organisms, machines, societies, and others. The cybernetic instinct rushes many visionaries to profound structural insights but also to overly determined design decisions. The circuitry of a computer chip and the neural networks of a mind do not resemble each other, although cybernetics earns its keep by finding usable analogs between them. This cybernetic system analog instinct—to design in beautiful symmetry where not necessary—helps to explain the consistent hierarchically decentralized design of all Soviet national network projects. They were designed to resemble the national economy as it appeared in principle, not as it worked in practice. To quote the secretary of the history of the Central Economic Mathematical Institute, a collegial institute of Glushkov's Institute of Cybernetics, the decision was made to "build the country's unified net hierarchically—just as the economy was planned in those days."[20]

In other words, Kitov, Glushkov, Fedorenko, and others followed the cybernetic integration of machines and biology to its design conclusions. Like Kharkevich's design, Glushkov's OGAS and other Soviet economic cyberneticists insisted that the Soviet economy, as a national body, needed a central information processor, administrator, and brain. They were not alone in modeling national networks as a neural network in the early 1960s.[21] The U.S. network engineer Paul Baran envisioned the ARPANET as a distributed packet-switching network that was modeled in part after Warren McCulloch's vision of the brain.

Note the difference here: Soviet economic cyberneticists under Glushkov conceived of the national network as a match for a national economic body with the network as the nervous system complete with a central processing in Moscow, and the American model of distributed networking imagined the whole of the nationwide computer network after the dynamic structure of the brain itself, not the body. In the Soviet Union, the command economy resembled the body, with the economic planning apparatus as its nervous system and Moscow planners as the brain, and in the West, after the ARPANET was commercialized, there was no body outside of the brain itself: the whole national network of users made up the nationally distributed brain itself.

To reduce it to a simplistic cold war binary: cybernetic network entrepreneurs throughout the world had competing analogs for thinking about national networks. In America, the ARPANET was designed to resemble a brain of the nation because its visionaries first imagined the nation as a single distributed brain of users. In the Soviet Union, the OGAS was designed to resemble a nervous system for the nation because its visionaries first imagined the nation as a single incorporated body of workers. This Soviet analog between network and nervous system, far from determining the outcome of the network, also occurred in Project Cybersyn in the early 1970s in Chile. Its principal architect, the British cyberneticist Stafford Beer, sketched the socialist Salvador Allende's nation as a viable system that was based on the "human nervous system" analogized with a comprehensive firm or corporate organization—complete with executives in adaptive feedback loops with the national body of workers.[22]

In addition to taking the cybernetic brain-computer analogy to its logical extreme, Glushkov also sought tight structural analogies in the communication systems that connected technical and human machines. For example, he designed the programming language Analytic to resemble human speech: "We continued to develop it in accordance with the principles of progressively complex machine languages, to get closer to human language.... My goal was to be able to speak directly with the computer and issue commands in our language."[23] Like the relationship between neural, processor, and national economic networks, Glushkov's thoughts about scripting together natural language and computer programming rests on the assumption that there is nothing particularly natural about natural language and that computing coding (like his other conceptual innovations in macropiping processing, automata, and the paperless office) represented an extension of the calculable artifice already hard at work in human behavior.[24]

Each of his innovations sought to reframe and solve knotty local problems in terms that scaled to a larger global system that contained those problems and all those like them. In fact central to understanding Glushkov's life and work and his scalable vision for the OGAS is his unflagging intellectual commitment to what he called "practical universals"—the merging of mathematics and economics, the theoretical and the applied, the universal and the particular. He and his colleagues repeatedly insisted that three principles guided his life work—"the unity of theory and practice, the unity of distant and near goals, and the decentralization of responsibility."[25] He taught others that before putting a principle into action, they had to formulate it into a general model or rule in abstract mathematical terms and then test that rule practically, applying it to countless concrete examples—an imperative to act locally while thinking globally. When higher authorities handed six of his researchers seven discrete system problems, Glushkov insisted that the first step was to develop a universal language for modeling all discrete systems, a language by which they could then solve all seven problems simultaneously, as well as any more they could be given.[26] The OGAS in design and implementation followed suit: people at each step of the network—including factory-based control system, regional computer center, and national economic planning center—sought to solve short-term factory problems by developing a universal system for advancing Soviet socialism toward communism.

The OGAS, for Glushkov, was to be a national communication network, countless local paperless offices, and a dynamic management system that connected them—a global-local network. A proper economic reform, in his mind, must benefit the factory worker, the general secretary, and the whole populace. The OGAS sought to pole-vault socialism toward communism at the Hegelian level of historical progress and to usher in a better work life for the knowledge worker: in the command economy, everyone needed to work knowledgably with economic plans. The OGAS would grant both at once, automatically storing relevant digital files on every local actor while granting remote access anywhere else in the country. The origins of the ideas behind the OGAS computing network also point to a preexisting academic network, including the circulation of a 1955 Academy of Sciences proposal by Nemchinov to erect large but unconnected state computer centers in Moscow, Kiev, Novosibirsk, Riga, Kharkov, and other major cities. However, this proposal did not connect the computer centers but instead specified that they should be built to facilitate the local exchange and standardization of scientific and economic information. Kitov's Red Book letter in the fall of 1959, which included the initial proposal to network such computers together into one, was the next step. In fact, after Kitov was dismissed from the military, Glushkov hired him to serve as a scientific adviser and personal confidant to his projects. Their respective trust network grew so close that, two decades later, one of Kitov's sons and one of Glushkov's daughters wed, signifying, just as the close connections between Baran and McCulloch, that personal communication networks both precede and outlast national computing networks.[27]

Beginning in the early 1960s, Glushkov's detractors recognized the sweeping commitment to practical universals in this vision and colored it in different shades. As he exercised his penetrating ability to formulate and scale up or down any problem by the force of mathematical reason, Glushkov's vision of the socialist cybernetic future moved, in the estimation of researchers at the Central Economic-Mathematical Institute (CEMI) and liberal economists, in "romantic" and "quixotic" leaps. Even his colleagues admitted in interviews that at the grandest vision, the OGAS ambition had an almost "religious" or cosmological reach to it.[28] The modern reader should suspend incredulity at the scope of his theoretical scale until after observing the similar scale of technological ambitions at work elsewhere. The totalizing corporate missions of modern-day major data companies and the scope with which data are harvested by corporations and states share intellectual affinities with the all-inclusiveness of his or any global-local network vision. Glushkov was not alone in 1963 in proposing that the state should gather dossiers on every worker and economic actor in his nation.

By contrast, Glushkov's proponents, caught up in both the breadth and precision of his plans, too often overlooked the frequent criticism that no institutional environment could possibly be ready to do all that the OGAS sought to do. Glushkov also recognized that no practical effort, no matter how impressive, could ever satisfy both the local and global demands of making paperless the command economy, and many of his career efforts outside the OGAS Project focused, to his credit, on local projects, including the paperless office.[29] For the OGAS, however, because it was a matter of economic bureaucracy reform, he insisted on a comprehensive meaning of economic information: "since the object of control is not only equipment but also personnel, one must include [in the OGAS] all the information about new technical, technological, economic, and organizational ideas and projects that workers at a given enterprise have."[30] Far more than a shared file containing economic information, the OGAS presented itself as a real-time clearinghouse for information concerning individuals, projects, factories, enterprises, and industries. The network would continue to expand in scope, according to Glushkov, until it encompassed the whole of the Soviet economy as well as all workers, their activities, and their office space. At worst, the vision appears a totalizing and decentralized (not totalitarian) information capture of the workers and their work environment. At best, it appears to be an organization information upgrade that is fit for every large-scale corporation. Depending on how one weighs the values of individual privacy and organizational purpose, these two champion a particular universal ethical tension that occupies the modern media age.

Glushkov also foresaw (or rather projected) a hint of the financial future, although perhaps not the future he had hoped for. Because the socialist economy would be incrementally organized into a cybernetically balanced network of labor, production, and consumption inputs and outputs, Glushkov reasoned, there would remain no reason not to virtualize currency itself and make the exchange of funds take place by "electronic receipt." With the OGAS operational, there would no longer be need for hard currency. All economic exchanges would take place online. Following this line of thought, Glushkov included in his initial OGAS draft proposal a noteworthy provision to eliminate all paper currency, providing in its place wireless money transfers, or a "moneyless system of receipts" over the OGAS network.[31] Although modern readers may be tempted to see in his proposal a prototype of the modern-day ATM, e-commerce digital money transfers, PayPal, or BitCoin, Glushkov framed paperless money transfers in the politics of his time and place, calling it the fulfillment of a Marxian prophecy of a future Communist society without hard currency. Read backward in presentist terms, as historians are loath to do, the proposal, if realized, would have transformed the Soviet Union into, in Vladislav Zubok's phrase, "a computerized socialist utopia, the motherland of the Internet and also possibly the ATM."[32]

I maintain that the historical lesson is that whatever our present-day language and whatever the future imaginations of hard currency, the past brims with a variety of visionaries who thought about the future of money as virtual, when as history instructs, the dominant form of currency has already always been, since ancient Mesopotomia, the arithmetic matter of credit and debit—itself a form of expectant funds, or money transfers made virtual across time, not space.[33] After reviewing the proposal, Keldysh, then president of the Soviet Academy of Science and a major supporter of Glushkov, asked to meet with Glushkov privately and urged Glushkov to strike from his original OGAS proposal the recommendation of a networked society without hard currency out of fear that it would raise "unneeded emotions." He warned Glushkov that the reviewing Soviet administrators were so deeply attached to the advantages of hard currency that no reasoning or ideological commitment could persuade them to abandon it.[34] Glushkov conceded the point, and the Central Committee initially approved his project, pending further review.

Glushkov as a Pragmatic Administrator

In many ways, Glushkov, whatever his sweeping visions in cybernetic theory, proved to be a pragmatic administrator in practice. Because he understood practical administration, he also knew that the inevitable limitations of his theoretical ambitions were, in fact, an important part and consequence of his approach to problem solving with practical universals. If the Soviet administrative system worked informally behind the scenes, then so must he, especially if he wanted to help to rationalize or formalize that same system. Unlike some of the stillborn or short-lived cybernetic proposals noted earlier, the longevity of the OGAS as a potentially viable network project owes a debt to the tenacious and pragmatic administrative acumen that Glushkov and his colleagues displayed in navigating, managing, and alliance forging in the administrative base of the Soviet state between 1962 and 1983. As illustrated by the Komsomol letter incident and repeated by many of his colleagues, Glushkov was sensitive to the political nature of the OGAS proposals and, with his upper-echelon supporters, strategically planned every step of coalition building around every part of his proposal: who would support what and why.[35] No naïve technocrat, he sought to shape and situate his proposal according to the governing logics of *blat* and personal politics.

His wireless currency proposal is a case in point. On Keldysh's advice, he promptly dropped the idea and turned his attention back to the practical universals that he would need to understand before he could integrate both macroeconomic designs and microlevel problems of the Soviet economy. Glushkov sought out and marinated himself in the practices of the actual command economy so that he would understand locally what he sought to reform universally. In the early 1960s, Glushkov received permission from Keldysh to observe how each of the constituent parts in the Soviet economy—factories, firms, collectivized farms of all types, and administrative organs like local, regional, and national planning committees—actually worked. His purpose was ethnographic—"to ask questions, or simply sit in the corner and watch how they work: what he decides, how he decides it, according to what principles, etc." Glushkov recalls, "And naturally I received permission to acquaint myself with any industrial object—corporations, organizations—that I wanted."[36] By 1963, Glushkov reported having visited and observed over a hundred such industrial sites and nearly a thousand over the next decade, including mines, *kolkhozes* (collective farms), *sovkhozes* (Soviet state farms), railways, an airport, higher control organs, and administrative organs at Gosplan (the Soviet ministry charged with planning the Soviet economy) and the Ministry of Finance. Glushkov claimed that "I may know the structure of the national economy better than anyone else: from the bottom up, I know the peculiarities of the existing controls system, the difficulties which occur, and the most important issues."[37]

At roughly the same time that the OGAS proposal was being reviewed by the Central Statistical Administration, Glushkov gained insights into the navigation of the informal complaint culture and the administrative mechanisms that were available for resolving them. Between 1966 and 1976, he served as a Kiev-based adviser for the Division of Clemency (*otdel' pomilovaniya*) for the prominent city of Kharkiv in eastern Ukraine. His behavior in this public function also provides a glimpse into his administrative behavior concerning top-secret projects such as the OGAS. In these archival materials, a pattern emerges. For each complaint case that he considered, he sent a formal letter and an informal letter. The first letter he sent to the complainant to offer his moral support but declare his likely inability to ease their situation, and the second letter he sent to the relevant supervisory institution pleading informally the strongest appropriate case on behalf of the complainant. Thus he resolved dozens of real-life conflicts within the actual social economy of formal appeals and complaints, including helping a grandmother campaign against alcoholism, speeding a mother's request for an apartment, acquitting a decorated war veteran convicted of unspecified crimes, and restoring to his studies a graduate student found guilty of "hooliganry" for being found in a "nonsober" condition.[38] At the same time that he was navigating this public trading zone between the superabundant conflicts of bureaucratic and real-life interests, he was developing the OGAS as a top-secret human-computer system proposal that would do the same—resolve informal conflicts at a national economic level. Next I look at how the informal behind-the-scenes work culture of these cyberneticists contextualizes this larger point.

"Cybertonia": From National Cyberculture to Local Counterculture

Glushkov's proposal to rationalize and automate the national economy in 1962 took shape just as his own institutional environment was being upgraded from a small computing center to a more ambitious formal setting of an academic institute, without losing its informal and, in after-work hours, almost countercultural work environment. In the early 1960s, his vision for reforming the command economy took on national ambitions at the same time that his own local institution entered national prominence. A glance at the institutional transition from Sergei Lebedev's laboratory in the valley of Feofania to Glushkov's Institute of Cybernetics will provide insights into how the local institutional culture of this particular transition animated both formal and informal attempts to imagine an alternate Soviet information society.

The formal history of the transition from computing center to academic institute is illustrious if not unusual. In the late 1940s and 1950s, Sergei Alexeyevich Lebedev gathered a small and extraordinarily talented group of electrical engineers into a computing laboratory in the valley of Feofania in the southern outskirts of Kiev, Ukraine. That small group brought into existence the MESM (*malaya electronicheskaya schetnaya mashina*, or the small electronic calculating machine, and predecessor to the mainframe workhorse BESM series), the first stored-memory electronic computer in Europe, arriving four years after von Neumann's UNIAC[39] (figure 4.4). In 1952, the first "large electronic computer," the BESM, or *bol'shaya electronicheskaya schetnaya mashina*, followed, and then a series of Soviet native mainframe computers—the M-20, the BESM-3M, BESM-4, M-220, M-222, and finally the BESM-6. Designed in 1966 and produced first in 1968, the impressive BESM-6 went into serial production and served in special-purpose computation centers and military computer networks for the next two decades. In 1962, under Glushkov's direction, Lebedev's laboratory was relocated a mile away to a separate campus facility of the future Institute of Cybernetics that

Figure 4.4
The MESM (small electronic calculating machine) and its team in the monastery near the cathedral in Theophania, 1952.

was known for a series of subsequent impressive achievements. Researchers at that facility developed the "Dnieper" computer series, which powered the base stations for Soviet cosmonaut flight south of Moscow while pressing the frontiers of Soviet information science and technology. The institute also is known for developing the mainframe and early microcomputers Mir and Promin and a range of research on economic cybernetics, medical cybernetics, artificial intelligence, optimization, and defense research. The projects included the first network project to digitize the entire command economy and their central project—the OGAS and its technical base EGSVT beginning in 1963. In all, the official histories convey the gravitas that one would expect from one of the elite teams of Soviet scientists.

A closer look at the local practices of these institutions, however, sheds a very different light on this moment of Soviet optimism. The years 1962 and 1963 marked the height of enthusiasm for a young, entrepreneurial, and surprisingly humorous and mischievous group of cyberneticists. Lebedev's laboratory was situated in a forest that was enchanted with Slavic legends. Overrun by songbirds, rabbits, mushrooms, and berries in the summer and haunted in the winter by rumors of wolves and Baba Yaga (the famous

witch of eastern European folklore), this forest served as a curiously natu-
ralistic cradle for Lebedev's MESM, which was then the emblem of the new
Soviet religion of rational scientific progress. In the center of an opening in
the woods stands St. Panteleimon's Cathedral (Panteleimonivs'kii sobor), a
high point of Russian revival ecclesiastical architecture since its construction in 1905 to 1912 (figures 4.5 and 4.6).

Nearby stands a two-story brick building that tells a story of a complicated intersection of faith, madness, murder, and science. Initially built as a
dormitory for Eastern Orthodox priests, the building was looted during the
1917 Russian revolution and converted into a psychiatric hospital. In 1941,
the Nazis murdered its patients and established it as a military hospital. In
1948, the badly damaged building was transferred to Lebedev's work on the
newest icon of Soviet atheism—that triumph of human rationality and creativity that was the automated computer. Six thousand vacuum tubes and
two years of astonishing effort later, Lebedev's team turned on the monster
calculating machine in 1950. A sense of collaborative, dedicated work ethic
lingered in the decades thereafter, and a sense of local autonomy that was
away from the watchful eyes of Moscow pervaded the area of Feofania.
Researchers who received housing nearby rarely chose to leave, even when
offered more prestigious positions. Informal play and even troublemak-
ing abounded. To the priests' chagrin today, engineers sometimes tested

Figure 4.5
St. Panteleimon's cathedral and monastery (left), which housed the MESM.

Figure 4.6
Park, pond, and forest in Feofania, the general setting for Sergei Lebedev's computing laboratory, late 1940s to 1950s.

controlled mechanical explosions in the magisterial monastery. Water fetched from a nearby well was used to extinguish the fires because the building where the first computer in Europe was built had no plumbing. After work, the mood lightened. Bus drivers were sent on wild goose chases through the forest, and juggling and ping-pong balls ricocheted down the hallways of offices and laboratories. On work breaks, volleyball and soccer games broke out, and after work, the researchers ran to swim in the nearby lake and to wander through the tall pines and oak trees of the surrounding forest. Lebedev and Glushkov are rumored to have drafted the organization of the Institute of Cybernetics, built three kilometers to the west, while strolling together through that forest.

When the Academy of Sciences appointed Glushkov to be the first director of the new Institute of Cybernetics, some of that informal spirit transferred to the new institution, in part thanks to a prolonged transition period during the 1960s in which the campus where the institute is currently housed was built. In the after-work hours and at holiday parties, the growing group of young institute researchers even imagined a humorous autonomous country of their own, "Cybertonia," a virtual country. The researchers, whose average age was roughly twenty-five, first christened this "fairytale [*skazochnaya*] land" during a New Year's Eve party in 1960. The joke snowballed. The fairytale land offered scientific seminars, lectures, films, and auctions mainly in the capital Kiev and an evening ball in the Ukrainian nationalist border city of L'vov, spinning off more and more activities (artwork, ballads, a short film, passports, currency), press releases, seminars, holiday and after-hour gatherings, community functions, and more parties.[40] The researchers at the Institute of Cybernetics were still several years away from occupying the Institute's future campus, which eventually included more than a dozen buildings along Glushkov Prospect in southwest Kiev (figure 4.7). From 1962 to 1970, the institute occupied a building at 4 Lysogorskaya Street several kilometers north, at an intersection with Nauka (Science) Street, an area famous for being featured in the science fiction of the Strugatskii brothers, who worked in the Institute of Physics a few blocks away[41] (figure 4.8).

In its informal practices, the Cybertonia society abounded in pranks, puns, and puzzling wit, recreating a country in the image of the autonomous Soviet automata. The collective issued fake stamped passports and

Институт кибернетики АН УССР, 1971—1977 гг.

Figure 4.7
Sketch of the Institute of Cybernetics campus, Prospect Academic Glushkov, 40, Kiev, 1970.

marriage certificates to the mostly male research staff and female administrative staff, authorized by the "Robot Council of Cybertonia."[42] (figures 4.9
and 4.10). Each passport packed mathematical equations into the blanks for
personal identification, accompanied by a national constitution and a map
of the future capital of "Cyber City" (Kybergrad). The workplace culture
at this prominent research institute embraced the joke as an ambiguous
means for letting a little steam off after work and, in their more ambitious
flights of imagination, envisioning a nation that was independent from the
Soviet Union. The blurring of reality and virtuality, work and play, science
and art was the point of "Cybertonia," a name that lives on in the title of
an academic journal recently begun by Glushkov's youngest daughter, Vera
Viktorevna Glushkova.[43] The Cybertonia constitution guaranteed the rights
to frivolity and humor complete with the faux-newspeak warning: "anyone
who disobeys the Robot will be stripped of their rights and cast out of the
country for 24 seconds" (figure 4.11). The map featured landmarks such as
"a Main Post Office and the Feedback Division (or Returned Communication)," or *Glavpochtamt y otdel obratnyi svazi*, a possible reference to Cybertonia as a self-contained system apart from the Soviet regime, as well as the
"Temple of the 12 Abends" (abnormal program ends, or software termina-
tions), or *Khram 12 avostov*, a near Russian homophone with "the Temple
of the Twelve Apostles." Currency was issued on the punch cards that were
used in analog computer memory storage.

Perhaps most boldly, the Cybertonia society hosted a saxophone-playing
robot mascot as a unveiled reference to jazz, an export of American global

Figure 4.8
Sketch of the Institute of Cybernetics building, Lysogarskaya 4, Kiev, 1966.

Figure 4.9
Cybertonia passport, 1965. (a) front, (b) back.

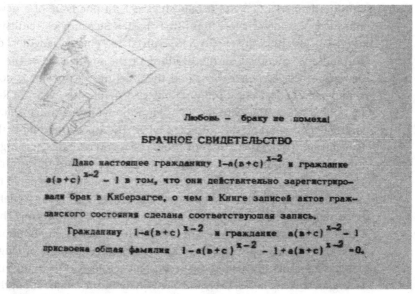

Figure 4.10
Cybertonia wedding certificate, 1965.

Figure 4.11
Constitution of the country of Cybertonia, about 1966.

culture (figure 4.12),[44] and it published at least one issue of a newspaper and made a comedic short film titled "Feofan Stepanovich serditsya" (figure 4.13). By 1966, its motto had evolved to "energy, laughter, dreams, and fantasy." Stamped on the headline of the single issue of the group's newspaper the *Evening Cyber* stood the greetings "s novyim kodom" (or "happy new code," a near homophone with "happy new year" in Russian). In 1968, a season ripe with revolt, a symposium of cybertonians published an irreverent report on the "complex cybernetic aspects of humor" that was issued from "Cyber City" in April 1969. The report contains nothing explicitly subversive but overflows with technocratic wit and sarcasm directed against Soviet authority figures. These merry pranksters compared the task of securing living quarters (a notorious challenge of everyday Soviet life) to hyperdimensional geometry and published "formal" reports on "theory of Graphs/Counts" (*teoriya grafov*, the royal title of *count* is a homophone with the word *graph* in Russian), a Jonathan Swift–like account of laughter at work as an underutilized national economic resource, odes to the virtues of Georgian soccer, cheese, beer, and a few chauvinistic laughs about the prospects of the feminization of science. Another report in 1965 bore the bold

Figure 4.12
Cybertonia logo: a robot playing jazz on a saxophone, about 1966.

Figure 4.13
Parody newsletter: *Vechernii Kiber* (*Evening Cyber*), 1966.

title "Executives Incognito: On Wanting to Remain Unknown, at Least to the Authorities."[45] Puns punctuated the technocratic discourse while quietly resisting power. These scientists sought in Cybertonia their own Cyberia away from Siberia, an escape from the great error of Khrushchev's age if not the great terror of Stalin's. Alas, Cybertonia never did grow to become, as the editors of its 1968 symposium had gleefully enthused, an "interplanetary congress." At some point between 1969 and 1970, as the Brezhnev doctrine compelled the Warsaw Pact to invade Czechoslovakia, "the entire idea of Cybertonia," as a participant recalled, "was buried by the pressure of the Party and government."[46]

The purpose of this snapshot into the informal lives of Soviet cyberneticists should be clear. In the forests of Feofania and in the virtual playground of Cybertonia, network entrepreneurs sought intellectual, political, and social autonomy, revelry, and even subtle informal protest from the oppressive regime that they served. Just as other cultures have demonstrated the rich connections between informal countercultures and cybercultures, lively network forums reproduced the cultural, institutional, and gendered mores of the Soviet 1960s, conceiving of a kind of privileged cybercommune of their own making.[47]

In the early 1960s—when Glushkov's ambitious plan to network, account for, and automate the nationwide command economy faced both partial formal approval and informal resistance from the top state authorities—his own local institution was undergoing significant institutional growth even as it was being told it must develop the EGSVTs before the OGAS network. In this fleeting period of optimism, the establishment and growth of the Institute of Cybernetics led to a form of institutional adolescence in which it exercised institutional ambitions on the national stage while informally and internally venting a kind of countercultural defiance against the state regime that governed it.

In fact, at the same time, 1962 to 1968, that Cybertonia was being celebrated during after-work hours, the Institute of Cybernetics was transitioning from a relatively small set of buildings near Theofania to a spacious campus a few kilometers to the southwest. It had enough modern buildings to house each major field of cybernetics with its own research department (except for Glushkov's "theoretical and economic cybernetics," which remained a department that preserves to this day the particular universal of Glushkov's merger of mathematics and economics).

CEMI and the OGAS Institutional Landscape in the 1960s

Glushkov's research institute was not alone in experiencing institutional growth in the early 1960s. Many prominent research institutes were established across the Soviet Union in the 1960s (Ukraine today has roughly 130 research institutes, and Russia has many more). Under the leadership of Aksel' Berg and the new president of the Soviet Academy of Sciences, Mstislav Keldysh, most of these pertained to cybernetic research. Those focusing on economic cybernetics included Viktor Glushkov's Institute of Cybernetics in Kiev and Nikolai Fedorenko's Central Economic-Mathematical Institute in Moscow.

These academic institutes were located in the capitals of the Soviet empire and under the umbrella of the Soviet Academy of Sciences. They functioned not as "islands of autonomy" (as may have been the case in the secret Siberian science city of Akademgorodok) but initially as contingent trading zones and eventually holding stations for enthusiastic young researchers who powered much of the early wave of Soviet cybernetic research growth throughout the 1960s. Prominent research institutes of all kinds sought to

establish a rolling range of connections, although they often were prohib-
ited from doing so in lasting ways. In many cases, the most crucial alliances
and associations for the survival and success of their core research projects
rested on currying productive relationships with the governing state ministries, not peer research institutes, whose areas of responsibility affected
their research missions. The CEMI in Moscow, for example, effectively
became an operations arm for Gosplan and other large ministries, and the
Institute of Cybernetics in Kiev maintained greater degrees of separation.
The history of how these institutional alliances unfolded is the short history of the OGAS Project and its undoing. Some attention will be paid in
the following sections to outlining the formation and deformation of the
alliances between economic cybernetic research institutes and Gosplan, the
Ministry of Finance, the Central Statistical Administration, and the Ministry of Defense.

In 1963, Glushkov's Institute of Cybernetics and another new powerful economics institute—the Central Economic-Mathematical Institute
(CEMI)—formed an alliance to advance the OGAS project, although the
seeds had been planted several years earlier. When Vasily Sergeevich Nemchinov—a senior economist-mathematician who was a strong advocate of
economic cybernetic reform and who had done much to introduce Kantorovich's linear modeling and input-output mathematical models into
Soviet economic planning—was proposing the CEMI in 1960, he initially
called it the Institute of Economic Cybernetics and devoted it to Glushkov's
main task of networking the national economy.[48] The founding of CEMI
receives a moment of attention, too, because both new institutes invested
hundreds of young researchers and dedicated funding streams into developing the OGAS project.

Before CEMI was an institute, it was a small laboratory in Moscow in
1958 called the Laboratory of Economical Mathematical Methods. Nemchinov appealed to the Ministry of Finances of the USSR by letter in January 1962, claiming that the transformation of the Soviet economy from
socialism to communism depended on "optimal plans for the nation's
economy."[49] By "plans" he had in mind the "optimal planning" of Kantorovich's linear programming as understood as both local microeconomic
modeling (which could be done on a standalone mainframe computer) and
a macroeconomic national infrastructure for processing the planned economy's commands by computer. Initially inspired by Kitov's failed 1959 Red
Book letter, Nemchinov appealed to the Ministry of Finance that "the modern mathematical methods and the means of mechanization and automation" were necessary to manage the complexity of the economy, invoking

Keldysh's call in 1962 for "the transformation of economics into an exact science in the full sense of the word."[50] Keldysh, the president of the Academy of Sciences of the USSR as of 1961, officially approved and promoted economic cybernetic research as a priority of the academy, underscoring that "the development of a theory of optimal planning and management to a unified mathematical model of national economy was one of the main directions of developments in modern economic science."[51]

Nemchinov also employed cold war rhetoric to provide a sense of urgency to his promotion of economic cybernetic methods as a means for governing a society, socialist and capitalist alike, noting the strong similarities between neoclassical econometrics in market economies and economic cybernetics in socialist command economies: "after World War II these methods were reopened in the West and were applied extremely widely to monopolistic government planning."[52] He then invoked a sort of cold war cybernetic economics gap, worrying that the Soviets had lagged behind the use of cybernetic methods "in the internal planning of the most developed capitalistic countries."[53]

Yuri N. Gavrilets joined Nemchinov's laboratory in 1959 and continues to work at CEMI to this day. A former rocket engineer, he was one among many military engineers who, like Kitov, was forced to pursue nonmilitary economic research after a youthful display of what was interpreted to be anti-Stalin activities in the late 1950s. According to an interview with him, the early efforts of CEMI to incorporate mathematical methods into optimal planning of the command economy openly sought to merge the best Marxist principles of social justice and planning with capitalist free-market equilibria.[54] Inflating the threat of capitalist cyberneticists, Nemchinov contended in his original proposal that "not a single scientific point" (which he later crossed out by pen and replaced with the word *center*) currently stood ready to "guide and coordinate research in the field [of economic cybernetics]" in all of the Soviet Union.[55] The Central Economic-Mathematical Institute, Nemchinov concluded, together with the OGAS and its associated mission of planning the economy by a cybernetic management network, would fill just such a gap.

Nemchinov drafted his CEMI proposal several days after First Secretary Nikita Khrushchev spoke at the Twenty-second Congress of the Communist Party on October 18, 1961. That "secret speech" is remembered today for denouncing the cult of personality, although Khrushchev also countered in it Stalin's bias against mathematical economics policy: "life itself requires a much higher class of scientific foundations and economic accounts from the planning and national economic leadership."[56] Cybernetics features frequently in Nemchinov's official explanation of CEMI's research tasks, such as "the wide application of cybernetics, electronic calculating machines and the regulating devices in production processes of industry, construction industry and transport, in scientific research, in the planning and project construction of calculations, in the sphere of accounting and management."[57]

In a letter dated November 17, 1961, a month after the Twenty-second Congress of the Communist Party, Nemchinov named four institute research directives that were dedicated to the network vision laid out in Kitov's 1959 letter and Glushkov's subsequent formulation of the OGAS:

1. The development of a unified system of planned economic information to improve planned information and documentation companies, including work on the application of modern calculating machines;
2. The development of algorithms for planned calculations based on a unified system of information;
3. Dynamic modeling for developing the national economy; and
4. Mathematical work for constructing a unified, centralized national economic plan, which would develop "the communist form of self-government of the production units, the optimal composition of general governmental interests, every company, and every worker."[58]

These quotations came from his earlier discussion of Lenin-Marxist rhetoric for economic planning. By wedding the cybernetic and Marxist-Leninist rhetoric of self-governing economies, Nemchinov sought to propose "economic cybernetics" and its plausibly nonsocialist "dynamic models of balancing capital investment" in the ideologically most acceptable light.[59] CEMI—in Nemchinov and his superiors' original vision—was set to become a powerhouse intellectual engine for driving a cybernetic vision of the networked national economy.

In late 1962, after receiving preliminary confirmation that CEMI would be established, Nemchinov, then age sixty-eight, grew too sick to continue his work and transferred the directorship of the Institute to the young academician Nikolai Fedorenko. Nemchinov died November 5, 1964, at the age of seventy. Had Nemchniov not grown ill, it is likely he, not Nikolai Fedorenko, would have emerged as Glushkov's first and strongest ally in Moscow.

Fedorenko and Glushkov: A Partnership Pulled Apart

In the beginning, Nikolai Prokof'evich Fedorenko proved a valuable colleague, confidante, and foil for establishing Glushkov's OGAS Project (figure 4.14). In 1962 and 1963, both cyberneticists were appointed the first directors of brand-new and prestigious academic institutes—Fedorenko's CEMI in Moscow and Glushkov's Institute of Cybernetics (IK) in Kiev—that, under their directorship and their shared vision of networking the national economy, led the Soviet Academy of Sciences in economic cybernetics. At the start in 1963, this dynamic duo of rising young academicians seemed destined to follow parallel paths to greatness while salvaging the failing Soviet economy along the way—at least according to Aleksei Kosygin, then deputy chair of the Council of Ministers, who was supporting their initiatives at the same time that he was advancing the liberal economic reform. All lights appeared green, and in 1964, the funds began to pour into these institutions to shore up the alternative to the Kosygin-Liberman reforms. The personnel at the two institutes multiplied exponentially. In a few years, the Institute of Cybernetics' staff numbers grew from dozens to over two thousand, and the ranks at CEMI sprouted from its original fourteen researchers in Akademgorodok to over one thousand researchers and staff in Moscow. Most of those new employees were young researchers with bold ambitions and a distaste for the culture of totalitarian control in the 1940s and 1950s. Enthusiasm for decentralized economic reform met with central flows of funding. In the late 1960s, after construction work was complete, CEMI moved into a state-of-the-art, twenty-floor skyscraper in the desirable Cheremushki neighborhood in Moscow, and after a decade of transition in the 1960s, the Institute of Cybernetics occupied a well-equipped campus along the scenic southwest edges of Kiev (figure 4.15). At least for a moment in the heady transition of 1962 and 1963, the two institutes appeared ready to remake the Soviet economy together.

One of the systemic sources of institutional volatility in the Soviet knowledge base was the oversized influence that individual leaders, like CEOs in modern Western culture, played in navigating and mobilizing organizational pursuits. In this sense, institute directors, such as Nikolai Fedorenko, appear entrepreneurial in the almost conventional sense of organizational leaders who take risks, invest in them, and mitigate the consequences of those risks by creative institutional problem solving. A year after the founding of CEMI, Fedorenko reported to the Presidium of the Academy of Sciences, USSR, that thanks to "the Institute work on the creation of methods of optimal planning ... savings [in the sector of transportation] have already reached about half a billion rubles."[60]

CEMI, in its early years, was not bound by the institutional logic of path dependence. Compare, for example, the major research directives that Fedorenko lists in his yearly reports between 1964 and 1969. In the first annual report (1964), Fedorenko lists the following six research directives,

Figure 4.14
Nikolai Fedorenko, date unknown.

all of which concerned the building of an OGAS-related wide-area information network (note the second and fifth, in particular): (1) develop a
theory of optimal planning and management for a unified mathematical
model of national economy; (2) develop a unified system of economic
information; (3) standardize and algorithmize the planning and management processes; (4) develop mathematical methods for solving economic
problems; (5) design and create a unified state network of computer centers;
and (6) derive a specialized planning and management system based on
mathematical methods and computer technology. Five years later, by 1969,
that number had been pared down to three concerned with optimizing
and modeling microeconomic problems. The network initiative had disappeared from its Moscow initiative.

In other words, by 1969, the year that the U.S. ARPANET went online,
CEMI was no longer actively pursuing any unified computer network projects. As a RAND analyst noted in 1971:

The most conspicuous feature of the latest version [of CEMI's research directives] is
the absence of any reference to the unified state network of computer centers. Also
missing is the proposed system of economic information. The projects, representing

Figure 4.15
Central Economic Mathematical Institute in Moscow, with Mobius strip statue, 2008.

research on the methodology of economic analysis and organization of new operational systems, are replaced by work on economic projects, a much less innovative and more conventional activity.[61]

In the same years, Fedorenko's CEMI had drifted from the OGAS Project and grew to nearly forty times the size of Nemchinov's original laboratory. At the beginning of the 1960s, the average age of its full-time faculty was about twenty-six years old; ten years later in 1973, when the institute surpassed one thousand employees, the average age tallied in at thirty-four.[62] According to Gavrilets, a lifelong faculty member at CEMI, the institute began as a lively and energetic place for critical and enthusiastic young economic researchers.[63] Although an increase of eight years in the average age of staff members over a decade probably reflects natural aging, CEMI's workforce was still relatively young and energetic and conditioned to believe they had the support to do anything. As a result, CEMI was not constrained by any formal agreements, as OGAS campaigners might have sometimes wished, to pursue OGAS and its associated network projects.

It might be that CEMI's eventual abandonment of the OGAS Project contributed to the failure of the USSR to reform its economic situation.

The shortcomings of these technocratic economic reforms were due both to the complexity of the reforms as well as the more foundational ad hoc complexity of the ministerial networks that were scrambling for funds in the first place. CEMI chose to devote its funding to microeconomic mathematical modeling of the economy (not the national networking of the economy) because its success as an institute depended on its iterative navigation and securing of state-approved funding. Instead of committing to particular projects (as Glushkov's Institute did, in part thanks to his personal leadership) or requesting and receiving funding to conduct basic, unspecified research (as was common in both Soviet and U.S. military spheres), CEMI had to defend and justify tens of millions of rubles in expenditures for specified civilian-political purposes. Fedorenko reasonably found linear programming and modeling optimal microeconomic interactions (with what he called the SOFE method) to be a more sustainable and less politically fraught task than networking the economy.

Funding of all sorts was earmarked for certain purposes, dependent on budgetary categories, constrained by values set in institutional history and shaped by practice, influenced by industry best practices, marked by gift-giver sources, and saturated in the politics of negotiation and expectation.[64] As the German sociologist and philosopher George Simmel maintained in his classic work on money, economic value is as much a matter for the philosopher and sociologist who debate orders of evaluation (or the realm of the study of value) as it is for the accountants, for whom the key interest is the measuring of monetary value based not on value itself but on the likeness or exchangeability of value.[65] Fedorenko worked out the research directives for his explicitly civilian research institute in a decentralized funding environment where economic value was subject not to a flat marketplace but a hierarchy of state interests. As the beneficiary of such interests, CEMI was free to redirect its research directives (in this case, away from the OGAS Project) and was constrained to justify those civilian research directives in politically acceptable terms (in this case, toward microeconomic modeling). The net effect of the decentralized funding environment for civilian projects, especially during the political freeze under Brezhnev, for the Soviet network institutional landscape was to redirect research toward more politically conservative agendas.

This conclusion might appear backward at first because CEMI's choice to focus on microeconomic modeling arguably shares more with liberal economic (or neoliberal) calculation of value and the OGAS Project appears to be a relatively conservative attempt to use technology to reaffirm and rationalize the (decentralized) hierarchical command economy structure.

However, given the disconnect between practice and principal, the OGAS Project appears both more philosophically bold and practically far-fetched of the two economic cybernetic approaches.

Another barrier—the same that Kitov encountered in his show trial—was the wall between civilian and military economies, and it began to strain the hopes for an economic network. In the spring of 1965, Fedorenko and Glushkov approached the Ministry of Defense to discuss the possibility of joining military network initiatives with their own OGAS dreams. Both Glushkov and Fedorenko's institutes were developing technically compatible, top-down, large-scale computer networks projects—and as Kitov had pointed out in 1959, the Soviet military already had several in operation.[66] The military networks were hierarchical and decentralized, loosely designed after the U.S. SAGE computerized air defense system, the first large-scale computerized command-and-control system in the world. And so with Kitov's Red Book show trial in mind, Fedorenko and Glushkov met with Defense Minister Bagramyan to discuss the matter. After an hour discussion in which Glushkov and Fedorenko did most of the talking, the Minister of Defense Bagramyan replied, according to Fedorenko's memoirs, with the following:

You are good men, and you are doing right by concerning yourselves with the economy of the people's money. But I cannot help you.... My friends, the state gives me as much money as I ask for to build the technical basis [of the network]. As far as I understand, they give you nothing. If I were to cooperate with you, they would give money to neither me nor you, since there is the opinion that economics is a scab on the healthy body of the governmental mechanism for planning and management.[67]

In Bagramyan's notion that "economics is a scab on the healthy body" of the Soviet state, we encounter a conflict of organizational self-interest. The Soviet military, which was the single greatest benefactor of the Soviet command economy, refused to cooperate with a civilian cybernetic project because of the prevailing disdain for the very economic management techniques that the cyberneticists were hoping to reform. This denial of a request for cooperation is an example of the unregulated freedom that the minister enjoyed when he acted in what he felt was his institution's best interest. This organizational dissonance repeatedly overwhelmed Glushkov's and others' attempts at systemwide collaborative reform.

After this encounter, it is unclear how far, if at all, Fedorenko pursued funding or collaboration with Glushkov's OGAS. As in the case of Paul Baran at RAND, funding decisions clearly favored defense rationales. The Minister of Defense was free not to cooperate with anyone because top-secret military missions enjoyed competitive advantages over other projects. This also meant that funding approval depended not on the will of top Party officials but rather on peer and lateral coalition building among organizations that were both cooperating and also competing for limited funding and influence in the Soviet state. This contradictory institutional space, where entrepreneurs seek to leverage organizational dissonance, exemplifies what I mean by *heterarchy*. *Heterarchy* describes the presence of ambiguities that result from competing formal regimes of evaluation, and entrepreneurs are those who trade on those ambiguities.[68] As a case in point, the Politburo claimed to oversee the goals of both the Ministry of Defense and Glushkov's and Fedorenko's institutes, and yet the Minister of Defense operated within heterarchical power structure that gave it no reason to recognize the Politburo's evaluation of the OGAS. To have done so would have questioned the necessity of the Ministry of Defense's own access to massive funding from the Politburo. Ministries, free of any single centralized operational logic that might be capable of legislating coopera- tion top-down, were free to not cooperate. They were also free to shut out peer-competitor institutions.

By the late 1960s, CEMI under Fedorenko's leadership had abandoned the OGAS and EGSVTs national network project to refocus efforts on the microlevel linear modeling of Soviet factories and enterprises. Fedorenko, a former chemist who was accustomed to microanalytic scales, claimed that CEMI's contributions to analyzing the national economy had better chances when applied to smaller, more manageable local scales, which his institute developed into the optimal mathematical planning method known as SOFE (System of Optimal Functioning of the Economy). In his memoirs, Fedorenko admits that the number of successful macrolevel eco- nomic analyses that CEMI produced in three decades "could be counted on one hand." In contrast, in the tally of firm-level analyses or smaller, Fedorenko counted hundreds of successes over several decades of work.[69]

A closer look at CEMI's stepwise separation from Glushkov's OGAS Proj- ect in the 1960s sheds some light on the negotiated compromises and qual- ities possessed by entrepreneurs like Fedorenko in the Soviet knowledge base. CEMI, under Fedorenko, went onto pioneer microeconomic modeling across the nation. In 1964, CEMI opened a branch in Tallinn, Estonia, and in 1967, a branch in St. Petersburg. In the 1966 preparations for the celebra- tion of the fifty-year anniversary of the Soviet regime, Fedorenko described the EGSVTs (network) project in glowing if slightly scaled-down terms: "An important direction of CEMI's research is the development and creation of a unified state network. This network should consist of three levels: a main computational center, a few dozen prominent computational centers, and a lower network. Such a structure will allow flexible information accounting and both operational management of the industry according to territories and the organization of planning accounts according to topics."[70]

After 1967, CEMI internal documents stop mentioning any network projects, whether OGAS or EGSVTs. What began as a small laboratory devoted to a wide-ranging civilian-use network for the management of the economy became, as a RAND report later called it, an "operational support agency" for the Gosplan.[71] Today CEMI is remembered for spearheading optimal planning methods with computerized and mathematical models in the Soviet socialist economy. As its website proclaims with silent hindsight on its early network ambitions, "When the Institute was founded in 1963, its main goal was to elaborate the theory of optimal management of the economy, applying mathematical methods and the use of computers to the task of practical planning."[72]

Like most rivalries, the subsequent rivalry between these two peer institutes, CEMI and IK, developed out of, in Freud's phrase, a narcissism of petty differences. After having their original OGAS mission tabled, both resorted to developing from the bottom up microlevel, factory-level economic planning. Even today, CEMI continues to pursue enterprise-level economic modeling, and IK continues to develop automated systems of management (ASUs) for individual enterprises. Fedorenko reported having improved hundreds of factory-level flow models every decade, and Glushkov claimed to have established ASUs in Ukraine, St. Petersburg, and beyond. Despite these successes, Glushkov in 1975 observed that humans entering "half-truths" were hampering automated control systems so that "we find ourselves somewhere between confusion and a search for scapegoats."[73] The few ASUs that were implemented fell flat, as well, according to the émigré mathematical economist Aron Katsenelinboigen, who reported that ASUs had little to no effect and sometimes even negative effects due to the expenses of installation. Managers, who were often older and wary of being replaced, often lacked the capacity to become familiar with, let alone master, the economic-mathematical methods that the ASU required.[74] As Glushkov later noted in *Pravda*, one automatic control system was dismantled and sold because it "impartially pointed out management's blunders and omissions."[75] What began as an alliance in the early 1960s around a network became a rivalry after the 1970s when cybernetic institutes disagreed over the relevance and proper role of the computer in economic planning. As these sections illustrate, the tensions resulted not from the roles of computing networks and information technology but rather a series of serious administrative, institutional, personal, political, policy, and social problems.

Management Missteps: "Supervision" and the Separation of the OGAS and the EGSVTs

In 1962, after Keldysh advised Glushkov to submit the OGAS proposal to the heads of the Communist Party without the moneyless payment system, Glushkov, backed by Fedorenko and others, submitted his original OGAS proposal for a chain of reviews by a number of Soviet government agencies. As a result, a commission was formed to review his proposal, which received preliminary approval, and in 1963, it arrived at the desk of the Party Central Committee and the Council of Ministers. At this point, Glushkov, Fedorenko, the chair of the Central Statistical Administration V. N. Starovsky, the first deputy minister of communication A. I. Sergeichuk, the vice minister of finance, and others gathered together as a commission to discuss and review the proposal and several thousand pages of associated materials. For months in 1963, the commission met and discussed the details of Glushkov's proposal, and each member tried to object to and reject specific measures in it. Despite the proposal's considerable political support to this point, including review by the Politburo and the Central Committee, the result was support for a technical computer network but not economic reform. For a period of time, the shell of the OGAS Project was approved for "finalization" at the hands of the Central Statistical Administration, and the heart of the OGAS economic reform was postponed until future review.

Thus, a technical network project—the EGSVTs—was born, and the automated management of the OGAS was put on hold. The technical network was deemed the Unified State Network of Computing Centers (EGSVTs, for *edinnay gosudarstvennaya set' vyicheslitel'nikh tsentrov*), and in response, the committees issued a joint decree titled "On Improving the Supervision of Work on the Introduction of Computer Technology and Automated Management Systems into the National Economy."[76]

The presence of the word *supervision* in the decree title here is telling. The government agreed to improve the supervision of the automated management of the economy, not management itself, which the top Soviet leaders recognized must be left to the distinctly not automated human bureaucracy of state employees and planners. In particular, the officials charged with approving the OGAS stumbled over Glushkov's distinction between a system that would make executive commands, which they feared, and a system that could command information about those commands—or the economic metadata. When faced with the possibility of controlling all economic information, the commission reviewers concluded, given the already tumultuous economic supervision in the early 1960s, that a national economic network could supervise, not directly manage, the command economy.

Similar to Kitov's Red Book show trial, the official rationale for the initial decision to strip the OGAS of any capacity to reform the planning process itself came with a justification that did not quite match the action. In this case, the Central Committee denied the automated management portion of the OGAS proposal due to what they deemed (not without contradiction) to be the inefficiency of rational management systems. In practice, the Committee apparently denied the request out of a fear that Glushkov's OGAS would strip its own unsanctioned informal control over economic power. Commission members who supported the OGAS also worried that even with top-level support, midlevel administrators would sabotage OGAS's efforts to rationalize their management powers. The initial 1963 decision to postpone the capacity of OGAS to reform economic planning took place as Khrushchev was falling out of power and limited Kosygin-Liberman liberal economic reforms were being introduced. The submarining of both reforms highlights the contradictions that faced the commanding heights of the Soviet state. No matter how obvious it was that the mismanagement of the command economy drove the state's economic woes, the state could approve no major reform without a sweeping revolution in how it managed itself.

Glushkov learned his lesson from the 1963 commission experience and scaled back and reframed his work on networking the command economy from direct management to indirect information supervision. Beginning in 1963, he publicly repeated that "the OGAS does not command the economy, rather it commands the flows of information *about* the state of the economy," although in theory and practice, Glushkov grasped the inherent politics of recordkeeping.[77] There are good reasons to doubt this position as a political compromise. As Glushkov theorized elsewhere, (1) a strict divide between data and metadata functions denies the basic cybernetic proposition of feedback loops that ensure that metadata observation is never influence-neutral, and (2) no organizational reform can ever be divorced from its political implications. Whether for political protection or otherwise, the OGAS team, not unlike other information omnivore projects, sought to ease its critics' concerns by asserting that it would traffic merely in metadata.

In subsequent OGAS preparatory proposals, Glushkov reframed the barebones EGSVTs network not as a matter of direct economic management but rather as a support for information management related to the national economy. This was so even though, in practice, the network he proposed also advanced data exchange and communication across the local and national levels. In the 1960s, Glushkov and his team tried at least two times to propose to Party leaders a technical EGSVTs network—first as an all-nation network (in 1963) and later as a regional network in Ukraine (in 1967).[78] A metadata management view informed both proposals: "To organize information flows on the national scale," as Glushkov once put it, "one needs to centralize interagency management of all information banks and computer centers, not the management of the economy."[79] In reorienting his claim from the politically entrenched national economy itself to the supposedly neutral territory of information banks, technical networks, and data clearinghouses, Glushkov adopted the abiding belief that was common among cybernetics and many digital technologist heirs in the neutral politics of code. Nonetheless, he proclaimed his task to be "not only scientific and technical, but also political," espousing the recurrent and troublesome idea that the politics of computation and technology are somehow more neutral than other politics.[80]

Instead of imagining a future communism arising out of exchanges ordered by an automated network, Glushkov envisioned the revised OGAS Project as the means whereby human planners might process accurate information about the economy via a national computer network. The Soviet computer network, like similar computing projects elsewhere in the 1970s, appeared foremost to be a "public utility" and a mass medium for serving information over great distance.[81] (Computers too were mass media in the age of mass media.) This revised model proved durable politically in part because it came with the added efficiency of promises of liberal pricing reforms, while capitulating to the more pragmatic demands of reforming an economic planning administration staffed with self-interested humans. Moreover, the revised emphasis on having OGAS manage the supposedly immaterial information about economic interactions (rather than command the actual economic planning) also proved a salient political hedge for the defense of the project going forward. Although striking near the heart of the state communist project's goal to transform the material well-being of every citizen, the OGAS defenders publicly defended their reform ambitions as merely immaterial and informational, even as the design analog and cybernetic philosophy quietly espoused the more fundamental fact that every information reform is also an organizational and thus social reform. This convenient rhetorical distinction holds in later developments of the OGAS Project, including Glushkov's emphasis on "paperless informatics" as a kind of successor to cybernetics as a theoretical vocabulary for the emerging socialist information society.

Conclusion

Despite the tensions outlined above, the initial 1964 decision to downgrade the OGAS from a full-service technocratic economic reform to an EGSVTs technical network was sensible from the point of view of rational state administration. The Soviet state was in a period of political and economic transition from Khrushchev to Brezhnev, so it was not yet ready to implement an economic reform like the OGAS. Its restructuring of the information infrastructure of the command economy was so global that it risked becoming a fully interactive networked political economy that was run by remote-access data exchange and communication. In contrast, the Liberman-Kosygin reforms invoked the scalable introduction of new accounting profit measures in select enterprises and factories that, as the liberal economists stressed, would cost no more than the stroke of a pen. In comparison, the OGAS Project was too big to begin.

So as Kosygin began to implement the profit measure reforms in 1965, the OGAS proposal suffered serious delays and was passed over for institutional review for "finalization" by the Central Statistical Administration, which was directed by one of the most outspoken opponents of the OGAS Project on the commission, Vladimir Nikonovich Starovsky. Starovsky had written to the chair of the Council of Ministers, K. N. Rudnev, as early as November 1963 that he could not support the OGAS proposal because it conflicted with the Central Statistical Administration (CSA) mandate to oversee statistical matters, noting "a basic unified state network, in the opinion of the CSA, should be the extant network of machine stations and factories" already under its supervision.[82] Starovsky's opposition would adjust but never reverse. In retrospect, Glushkov singled out Starovsky's resistance: "later, when the fate of the OGAS was being decided, the leader of the Central Statistical Administration spoke against the project so much more furiously than anyone else that he did much to seal its sad fate."[83]

Still, the Central Committee did not reject the proposal and mandated that the CSA would be in charge of finalizing the project. Stuck between a rock and a hard place, Starovsky chose to resist by other means: the CSA submitted the OGAS proposal for finalization review by sending it off to its most remote regional departments in Archangelsk and Karakalpak in Siberia, where it underwent several years of what OGAS supporters recall as a series of interminable and often incoherent feasibility reviews and often nonsensical dataflow testing. The specific missteps of the Siberian CSA review—such as arbitrarily declaring that after accounting for overhead hardware costs, calculating economic problems by computer would be on average ten times more expensive than calculating the same problems by hand—are symptomatic of the information organization problem that the OGAS sought to resolve and rationalize. In command economies, the more information involved in planning, often the more opaque or meaningless that information becomes. (It was never clear why computing by machines should be ten times more expensive per calculation than by hand, and yet the number stood with the force of administrative fiat.) Starovsky was concerned that the OGAS would wrest from the CSA its central task of gathering statistics for managing the command economy, and so by introducing and inventing dubious feasibility information about an already uncertain OGAS Project, he effectively stalled the economic reform portion of the proposal from making progress at the national level through the rest of the 1960s.

The global-local character of Glushkov's decentralized design was part of the genius of the project and also illustrates how the OGAS Project continuously threatened the economic bureaucracy that it was meant to reform and serve. Glushkov's decentralized design of rational management could work only if it was implemented top-down with the support of a centralized administration, such as the CSA. But no centralized administration could be found to support it because, in practice, centralized administrations in the civilian sector benefited from not behaving like centralized administrations. Here we can begin to see the political paradox that the OGAS encountered—one that was manifest in the tensions between the formal master plan and the informal practices of the Soviet system and also in the life and work of one of the Soviet master mergers of theory and practice. Glushkov was aware that it was in the self-interest of institutions in the Soviet knowledge base to resist the OGAS. Despite having unparalleled insights into how official and shadow economies worked, Glushkov had no other choice but to model the OGAS network after the formal command economic model, not after economic behavior.

Part of that design choice is an intellectual consequence of the cybernetic instinct to analogize the social and technical into structurally similar information systems, such as the command economy and its national network. But a great portion of the design choice to model the OGAS after the formal command economy follows from political necessity. Consider this contradiction of political practice. To implement a fully decentralized reform to the economy, top political support needed to be secured to implement the reform systematically. The full decentralizing reform first had to be implemented with centralized systematic approval of the top. To gain the support of those top central authorities, the reform design had to conform to the publicly approved and ideologically acceptable principles of the current economic organization, which means that the OGAS design had to map onto the pyramid structure of economic planning in principle. So far, there is no contradiction because the short story of the tumultuous history of Soviet economic reforms is effectively one of top leaders who variously attempt to reaffirm their own hierarchical control, no matter how decentralized.

The contradiction lies in the practical need for the reform in the first place. The need for decentralized economic reforms follows from the fact that, as discussed, the command economy in practice never functioned in a strictly centralized manner. OGAS supporters sought to transform the economy into a decentralized hierarchy, but the economy, whose leaders publicly defended their positions in a centralized hierarchy, never behaved as a strict hierarchy because those leaders and their supporting personnel benefitted by the informal economy of favors and heterarchical connections. Many of those in the economic bureaucracy resisted the OGAS because although it purported to support the formal power structure that legitimated their positions, it also threatened to strip their institutions of the thing that justified their existence—the need to manage the command economy in the first place. The OGAS, if effective, would strip those positions of what made them informally beneficial to hold—the potential for corruption and personal gain and power. The organizational dissonance coursing throughout the command economy both motivated the reform and caused this initial frustration. With no other choice but to appeal to the top, the OGAS Project was stranded by the potential adopters of its decentralized design (the CSA in the late 1960s and other institutional entanglements in the 1970s and 1980s) because the project sought to resolve the conflicts of interest in the command economy that kept its own bureaucracy from resembling in practice the pyramid of political power that it had to appeal to.[84]

OGAS did not meet its end at the hands of stalled feasibility reports by the Central Statistical Administration between 1964 and 1969, however. During these years, Glushkov, among others, built considerable political support for developing the technical network of the EGSVTs. The late

1960s were a helpful preparatory period for building and securing political alliances that a small group of cyberneticists and network entrepreneurs attempted to form in the 1970s. This period was spent quietly and carefully working within the administrative heterarchy to secure political support. To a surprising degree, Glushkov succeeded in doing so at the upper echelons of Soviet power. Two top-ranking powerbrokers offered relatively unwavering support of the OGAS Project in the late 1960s. First was Aleksei Kosygin, who was effectively second only to Brezhnev in civilian matters. He was initially chair of the State Planning Committee (Gosplan) and first deputy chair of the Council of Ministers under Nikita Khrushchev (1959–1964) and rose under Brezhnev to become premier of the Soviet Union. As already noted, when Kosygin's initial profitability reforms in 1965 were met with fierce resistance from the economic administration and effectively stalled, Kosygin turned to the OGAS as the next best approach. Second was Dmitry F. Ustinov, who was a prominent military leader and manager who, just before helping ousting Khrushchev in 1964, served as first deputy premier with control over the civilian economy. In addition to being a career member of the Central Committee beginning in 1952, Ustinov ruled as the leading defense minister of the Soviet Union from 1976 to 1984.

Not long after the commission decided to postpone the OGAS in 1964, Petro Shelest—the first secretary of the Ukrainian Communist Party—called Glushkov to persuade him to cease promoting the OGAS and return to work (as Fedorenko in Moscow had already begun to do) on local or microeconomic systems. Gluskhov and his team complied with Shelest's commands and turned their attention back to developing local and regional computing centers that might later be connected by telephone and telegraph cables (figure 4.16). Soon after, Dmitry Ustinov countermanded Shelest's wishes, at least for the military: Ustinov, who was on his way to becoming minister of defense (1976–1984), invited Glushkov's team to build ASUs in test military factories.[85] Military support appears to have given the team the administrative license to advance the cause of computing technology and also to have ensured that their ASU work would not benefit or network the civilian economy.

In the 1970s, several civilian factories received ASUs under direction of the OGAS team. Most of these efforts were carried out from the bottom up, although Glushkov and his team at the Institute of Cybernetics continued to seek and occasionally secure top-level support in the 1970s only to see it dissolve in committees convened by intermediary ministries. For example, Glushkov, with the support of the director of the S. O. Petrovskiy television plant, successfully developed in two years local control systems such as the

МАШИННЫЙ ЗАЛ ГИВЦ

Figure 4.16
Inside an ASU: Machine Hall, State Institute of Computing Centers, unknown date.

L'viv System or Lviv MICS—an automated control system for streamlining the industrial processes in the Elektron television factory in L'viv, Ukraine. After completing the L'viv System, the team engineered a more complicated Kuntsevo system for planning and managing the resources of the Kuntsevo radio manufacturing plant in southwest Moscow.[86] The Institute of Cybernetics also proposed an industrywide network of ASUs in the industry-rich Donbass region of Ukraine (figures 4.17 and 4.18).

Not all installations went smoothly. One factory manager, Valentin Zgursky, senior technologist at a manufacturing plant, admitted that "when you brought the Universal Control Computer [a mainframe behind the ASU] to our plan for mass-production," Malinovsky recalls being told, "I did everything possible to make sure it would never succeed!"[87] Nevertheless, Zgursky eventually saw the value of the ASU and installed it (although his admission may have been the exception in the long run). Bolstered by some local successes on the edge of an empire in the late 1960s, Glushkov also repeatedly reminded anyone who would listen about the work that even a dozen or so local systems (ASUs) could do after they were connected into a single national network.

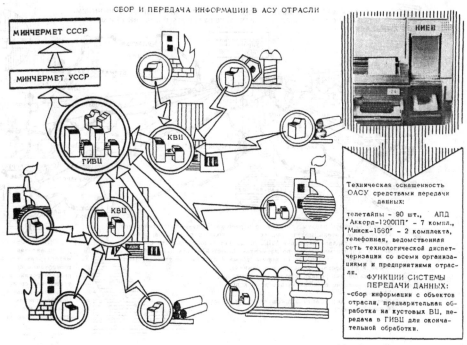

Figure 4.17
Diagram of an ASU network at the industry level, about 1969.

In the late 1960s, after these and other limited local successes, top leaders began to heed some of Glushkov's calls more carefully. Dmitry Ustinov commanded the heads of the military ministries to heed Glushkov's orders while he continued work on the L'viv System. After securing Ustinov's top brass support, Glushkov claimed that as early as the late 1960s, the automated management systems in factories throughout the empire provided the outline of what would become the OGAS: "it was planned from the very beginning that the whole system would apply across all spheres at once, so some rudiments of an all-state system were conceived"[88] (figure 4.19). After this chapter's discussion of the bold vision and rocky institutional landscape that supported the OGAS Project, the following chapter chronicles and comments on what happened when, in 1970, the Soviet centralized command decided to review the OGAS proposal to decentralize the economy by network in earnest.

In summary, in this chapter I have examined the OGAS design thinking that motivated Glushkov and his teams and the initial obstacles that

Figure 4.18
Map of the proposed ASU train industry in the Donbass, Ukraine, about 1969.

Figure 4.19
Viktor Glushkov giving a presentation on the ASU, about 1969.

they encountered. Against the bold vision of a networked electronic socialist future, a tangle of historical episodes frustrated the realization of that
vision. This chapter has offered a look at the institutional landscape and
alliances that formed and then dissolved between Nikolai Fedorenko's Central Economic-Mathematical Institute and Glushkov's Institute of Cybernetics, their heydays as the leaders of economic cybernetics and networked
cybernetic reform through the late 1960s, the informal work culture of the
Kiev-based cyberneticists in the 1960s, and early bureaucratic barriers that
slowed the advance of the OGAS Project in the Soviet military. Neither
the Ministry of Defense nor the liberal economists wanted to collaborate
and support the OGAS Project, perhaps because the country had endured
four turbulent years, from 1962 to 1966. In that time the Soviet Union
had agreed to pursue computer-aided economic reforms, come to the brink
of nuclear disaster in Cuba, forced out and replaced its general secretary,
founded and funded leading economic cybernetic institutes devoted to
building a national network plan, foregone approving the original proposal to introduce liberal profit reforms, and continued to fund the leading
economic-mathematical research institute in Moscow as it reoriented itself
away from its original network resolution to focus instead on less risky local
optimization and modeling problems. Topsy-turvy institutional behavior
in civilian matters was the rule, not the exception.

5 The Undoing of the OGAS, 1970 to 1989

No single actor could either make or undo the OGAS (All-State Automated System) Project. The hidden networks governing the Soviet state were far too complex and heterarchical to have had any single cause. (Most multiactor networks involve complexities that are impossible to express in linear form.) This chapter briefly outlines and analyzes the slow struggle over the political execution of the OGAS Project in the 1970s and its aftermath in the 1980s. The prolonged struggle and decline at the hands of various forces helps to reveal the complex heterarchical forces that governed the Soviet state and attempted to carry out economic and technological reforms. The commentary that follows speaks by analogy to modern observers who are concerned with attempts to reform complex political economic systems and also reflects on how attempts to create formal computer networks are sometimes thwarted by hidden social networks.

In this chapter, I chart the institutional apex, plateauing, and decline of the most ambitious attempt to provide the Soviet nation with its own form of networked socialism. The chapter begins by rehearsing the 1970 Politburo review of the OGAS proposal, the ministerial defiance and contingent institutional interests that extinguished its approval at the last minute, and the subsequent dozen years (1970–1982) of attempts by Glushkov and his team to revitalize state and then public interest in a networked socialist economy. The chapter then takes a detour through an unlikely case study before becoming reflecting on the central theme the military-civilian divide that separates hierarchical and heterarchical institutions in the Soviet Union. This case study examines how militarized strategic thinking—in the hands of one of the great Soviet chess masters—materialized into a stillborn attempt to plan the nation's political and economic strategies with early Soviet computer chess.

Ministry Mutiny

The strategic move that eventually stalemated the OGAS Project did not come from abroad. It came from within. By the summer of 1970, Glushkov, Ustinov, and others had mobilized enough support for a fresh review of the OGAS Project by the highest committees in the land. All signs, except one, suggested that the timing for an economic network renaissance was finally right. The Central Statistical Administration (CSA) could no longer delay its "finalization" review process for the OGAS proposal, which formally ended in 1966 but had lingered in approval limbo ever since. Simultaneously, the successful evidence of the local foundations of the EGSVTs (Unified State Network of Computing Centers) was gaining more and more support, especially as Party leaders searched for an untried approach to economic reform in the wake of the faltering Liberman reforms. By the time that Viktor Glushkov and Nikolai Fedorenko's partnership drifted into a rivalry over the wisdom of economic reform by macronetwork (Glushkov's OGAS) or micromodeling (Fedorenko's SOFE, or System of Optimal Functioning of the Economy), the EGSVTs had become such a promising project that established rivalries were reigniting over whose administration might best oversee its development and command the funding streams that came with it. By early in 1970, Vladimir Starovsky's Central Statistical Administration and Vasily Garbuzov's Ministry of Finance began to jockey for position to command the administration of the OGAS Project. These two powerful ministries began contending not just for the project but against one another in an effort to limit the competitor from securing massive funding.[1]

The most vocal opponent to the OGAS proposal in 1970 was also the man who officially had been charged with its care and finalization for the previous seven years. Vladimir Starovsky, the head of the Central Statistical Administration, "harshly objected to the whole project," Glushkov recalled in the late 1960s—out of opposition not to the economic reform but to the prospect that the Central Statistical Administration would have to cede control over some element of the governance of his administrative turf (economic statistics) to future OGAS directors. Starovsky rejected the remote-access portion of Glushkov's proposal (a precursor to "cloud computing"). If realized, the OGAS was going to provide access to information and processing power to any authenticated user anywhere on the network. Even though the permission hierarchy for authenticated users presumably could still reaffirm the strong hierarchical structure supporting his administration, Starovsky opposed what we now recognize as a cloud computing provision as being politically "unnecessary" because the Central Statistical

Administration was "organized by the initiative of Lenin" and already does everything that Lenin asked of it. Reversing Lenin's original question, "What is to be done?," Starovsky concluded that, because of Lenin, "Nothing needed to be done."[2]

From 1964 to 1970, as the CSA and Ministry of Finance were butting heads, another front of intellectual opposition arose against the OGAS Project from its own closest allies for economic reform—liberal economists. In 1964, a pivotal year for reform, Liberman, Belkin, Birman, and others were able to convince Kosygin that, in contrast with the nearly 20 billion rubles that the OGAS was predicted to cost, the cost of liberal economic reform would be "no more than the cost of the paper on which the resolution of the Council of Ministers would be printed."[3] Glushkov was caught unprepared for this counterattack, having already admitted to Kosygin that the whole network project would be net profitable but would prove to be more costly and complicated than the space and atomic programs combined. Nonetheless, the OGAS reform had the strategic advantage of not abandoning Marxist planning principles for liberal market ones and of promising to pay for itself quickly (and Glushkov foresaw a reimbursement of 5 billion rubles by the end of the next five-year plan).

The Day of Reckoning: October 1, 1970

Several factors led to the Politburo's review of Glushkov's OGAS proposal on October 1, 1970, which was the closest that the Soviet Union ever came to approving a national network of its own design. In the midst of a larger space and technology race, the unexpected revelation that the ARPANET—the first American civilian nationwide network—had gone online one year before, on October 29, 1969, suddenly hastened the search by top Party leaders for a viable local national network project. Knowing that the ARPANET was worrying Party leadership, Glushkov approached A. P. Kirilenko, then secretary of the Central Committee, to ask the committee to revisit the ideas in the previous proposal. Kirilenko welcomed the idea and asked Glushkov to "write down in detail what has to be done," said A. P. Kirilenko, "and we will create a commission." Glushkov wrote in reply: "The only thing I ask is not to create a commission. Commissions operate on the principle of subtraction of brains, not summation, and they can wreck any project."[4]

Nevertheless, Party leaders insisted on creating a commission. Glushkov declined to chair it, and so V. A. Kirillin, then chair of the State Committee for Science and Technology, was appointed as chair with Glushkov as his deputy. The oppressive Soviet interventions in Czechoslovakia sent a wave of recentralization, or rather antidecentralization, criticism through the state, and some Gosplan officials openly criticized EGSVTs proposals. The Politburo also felt pressed to consider and approve meaningful reform projects for the drafting of the Twenty-sixth All-Party Congress and the starting of the eighth five-year economic plan in 1971. As a result, the Politburo twice reviewed and approved without change Glushkov's OGAS wordings for the draft portion of the Twenty-sixth Congress. A preliminary meeting of the same review commission (which had dragged its feet since 1964) concluded in 1970 that the full OGAS, including the economic management part, should be approved for top-level review, although who would steer it after it was approved remained strategically unresolved. In particular, it was left unclear whether further "finalization" by the Central Statistical Administration would be required.

This time only one person on the review commission did not sign onto the newly revived OGAS proposal—the minister of finance, Vasily Garbuzov, who was the primary opponent to the CSA. Garbuzov refused to sign because he did not want the OGAS to fall under the control of his competitor institution, the Central Statistical Administration, whose director, Starovsky, also temporarily withdrew his support for the same reason several years earlier. Glushkov and his team deliberated over how to proceed. He did not want to submit his proposal to the Politburo for review if it lacked unanimous support, but he also knew that he could not resolve Garbuzov's concerns. Thus hedging its bets and hoping that the U.S. ARPANET would sway the Politburo into action, the commission (unofficially led by Glushkov) submitted the proposal for review.

That fateful gathering took place in Stalin's former office in the Kremlin. As Glushkov walked into the long, red-carpeted room, Kirillin, one of Glushkov's supporters in the Politburo, leaned over to whisper that something had happened but he did not know what. Before he could clarify, Glushkov noticed that something was out of place: the seats of the two most powerful men who should have been in the room were empty. General Secretary Brezhnev and his prime minister, Aleksei Kosygin, did not see eye to eye about many things, but on the matter of network economic reform in the fall of 1970, they appeared to be ready to make an uneasy truce. As it happened, Secretary General Brezhnev, who was a technocrat with an engineering background and was favorably inclined to sweeping technocratic solutions (especially those that disadvantaged orthodox economic planners), happened to be away for the day in Baku attending the fiftieth anniversary of the Soviet rule in Azerbaijan. Glushkov might have counted on Kosygin's support, but he too was away, pressing hands among the mourning crowds in Cairo at Gamal Abdel Nasser's funeral, who had died of a heart attack two days earlier. Both men—the first and second in command, including the economic reformer who was most likely to lobby for the OGAS Project—could not attend the fateful meeting because of calendar contingencies.

Despite these key empty seats, the meeting began well enough. Without Brezhnev and Kosygin in attendance, the meeting was conducted by the Stalinist-era Mikhail Suslov, who was famous for resisting radical changes as the "Chief Ideologue of the Communist Party" and a consummate behind-the-scenes operator with seats on both the Secretariat and the Politburo. Given this steely reputation, he began encouragingly by saying nothing against the proposal. Glushkov was then invited to speak, which he did briskly before responding to a series of questions to the apparent satisfaction of all involved. This went on for less than half an hour, until several higher-ups began to speak positively about the project. Baybakov, one of Kosygin's deputies, volunteered that if the Politiburo should make him head of the State Planning Committee (Gosplan), he would eliminate or merge three ministries so that staff could be found to support the OGAS Project. In this deft maneuver, Baybakov managed to relay Kosygin's enthusiasm and promote his own career. The Minister of Instrument Making, Automated Equipment, and Control Systems (Minpribor), K. N. Rudnev, had extolled the virtues of information technology in economic planning in 1963, signed the document, and commented off the record that the timing might be bad.[5] A chorus of voices countered these hesitations with unambiguous support of OGAS.

Just as it seemed that the committee might be nearing consensus approval, the minister of finance, Vasily Garbuzov, stood up. According to Glushkov:

[Garbuzov] entered the stage and addressed Mazurov, Kosygin's first assistant. He said that, well, he went to Minsk as directed, to examine the poultry farms. At the so-and-so farm, the workers designed a computing machine on their own. I laughed out loud. He shook a finger at me and said, "You, Glushkov, shouldn't laugh. We are discussing a serious issue." However, Suslov interrupted him: "Comrade Garbuzov, you are not the chairman here, and it's not up to you to control the proceedings of a Politburo hearing." He shrugged and self-confidently continued, "The machine can perform three programs—turns on music when the hen lays an egg, turns lights on and off, and so on. This increased egg production at the farm." So he suggested that first we should implement these machines at all the poultry farms in the Soviet Union and only then could we even begin thinking about silly projects like a nation-wide system.[6]

At that point, Garbuzov, who served as minister of finance for another fifteen years until his death in 1985, made a counterproposal. The OGAS should be released from the control of the Central Statistical Administration and put under the direction of a new institute that should develop (as the commission had insisted back in 1963 as no more than the EGSVTs barebones technical network) computers with lights that flash on and off. "Everything related to economics and the elaboration of mathematical models for the OGAS, etc.," Glushkov recalled, "was wiped off."[7] From a technical perspective, Garbuzov argued, the EGSVTs approach made political common sense. A technical network would avoid the minefield of economics, politics, and ideology without foreclosing the possibility of introducing relevant economic programming into that network in the future. This technical vision, Garbuzov argued, was the most risk-averse way forward.

Behind the veneer of Garbuzov's technical pragmatism lay a more self-interested motivation for this counterproposal. Having not been able to secure the OGAS for his own ministry, he preconditioned his technically reasonable counterproposal on the fact that a new institute should be developed to oversee the OGAS. If his ministry could not have the OGAS, then no other existing administrative entity should have it, he reasoned. After all, by what other way can a minister reduce the bureaucracy except by creating a new bureaucratic body to do so? By so specifying, Garbuzov sought to streamline the network development and submarine the chances that this competitor organization, the Central Statistical Administration, had of securing the massive funding streams and political gravity associated with commanding the management and automation of the command economy.

As the Politburo discussion ensued, the consensus slowly shifted from Glushkov's OGAS in favor of Garbuzov's EGSVTs counteroffer. At last Suslov intervened, concluding the discussion with executive authority: "Comrades, perhaps we are committing a mistake by not adopting the project fully, but it is such a revolutionary improvement that it will be hard for us to realize right now. Let us do it that way, and we will see later how to proceed." Suslov then asked what Glushkov thought, to which he responded pointedly, "Mikhail Andrevich, I can only say one thing: if we do not do [the full OGAS] now, then in the second half of the 1970s the Soviet economy will encounter such difficulties that we will have to return to this question regardless."[8]

Intrigue and unconfirmed speculation abound about how Garbuzov's Ministry of Finance managed to turn the Politburo against the OGAS that day. Prime Minister Kosygin, who probably would have pressed for a consensus in favor of the full OGAS, may even have chosen to attend the Nasser funeral to avoid having to cast a negative Politburo vote on the OGAS decision. Two years after the decision, in 1972, Glushkov heard rumors about the apparent backstory behind Garbuzov's counterproposal. Before the October 1 Politburo gathering, Garbuzov purportedly sought a private meeting with Prime Minister Kosýgin to convince him that if the CSA were allowed to govern the OGAS national project, the CSA would grow so powerful that it could wrest control over economic matters from Kosygin himself and the Council of Ministers, ceding it back to the Central Committee.[9]

If the OGAS was approved, the minister of finance argued to Kosygin, the Central Statistical Administration would surpass even Kosygin in economic power. Garbuzov almost certainly did not make this warning out of good will or concern for Kosygin's position. His ministry had done the most to undermine Kosygin's political reforms during the prior five years. Since 1965, the Ministry of Finance had informally refused to implement the Kosygin-Liberman reforms, thus encouraging discrediting criticism of the reforms before they could take full effect. The winning argument appeared to be a contradiction: Garbuzov contended that if Kosygin did not act to preserve the status quo, Garbuzov's competitor would strip Kosygin of the power to make economic reforms. Faced with that option and ceding the OGAS Project to the Ministry of Finance, Kosygin appeared stuck, although whether Kosygin actually believed Rudnev's argument does not matter compared to the result. From late 1970 until his retirement in 1980, Kosygin never moved to unmire the OGAS Project administratively.

Such were the moves and countermoves that were at work behind the administrative end game of the OGAS Project in the Politburo. A more generous reading upholds the possibility that Kosygin, the great liberal economic reformer, did not wish to yield to Garbuzov but nonetheless felt compelled to do so because a disgruntled Garbuzov and his ministry might sabotage any of Kosygin's future attempts to make economic reforms, whether or not the OGAS Project was governed by an independent administration. For Kosygin, the decision to neglect the OGAS could have been the best way forward in a lose-lose situation of mutually assured ministry mutiny between the CSA and the Ministry of Finance—short of risking his own power to make economic reforms. Because of the consequent ambiguities, every administrator had to engage in a form of entrepreneurial negotiation among their private plans, their competitors' plans, and the state plan. This tangled heterarchy at the top of the heap led every administrator with a stake in the decision into a competition with his neighbors. Historical contingency played a role, as well: perhaps there was to be no OGAS that day simply because two seats were vacated by a general secretary who had to attend a planned celebration and a prime minister who had to attend a funeral. At least Kosygin could wash his hands of having to make any top-level decision to advance the OGAS.

The OGAS Project was neither fully rejected nor approved. Instead, the Twenty-fourth Party Congress in April 1971 agreed with the Politburo decision that the ninth five-year plan (1971–1975) would establish some skeletal semblance of the OGAS, including 1,600 ASUs (automated systems of management); expand computer production by 2.6 times; and establish a technical network, the EGSVTs, across the nation. The EGSVTs, in this iteration, were to connect all higher-level branches and departments in the planning administration, develop regional networks, and connect and consolidate the regional networks to the higher-level network. The proposed details in 1971 were scaled back closer to the initial 1963 EGSTVs proposal levels—with twenty to thirty regional centers and the piecemeal incorporation of the national economy lurking in the background.

The OGAS Project in Repose

Having secured partial approval for the second time in a decade at the hands of a top-ranking commission but being no closer to his goal, Glushkov soldiered on in his commitment to introduce some kind of technocratic economic reform. Despite the authorities (the three words that he used to title his memoirs), Glushkov and his team installed ASUs (automated systems of management) in local factories with the hope of one day connecting them. Between 1970 and 1977, Glushkov and his team offered up a variety of decentralized network designs, although these proposals never satisfied a wide range of relevant parties.[10] A Ukrainian computing pioneer, Boris Malinovsky—who for his technical and historical achievements should be remembered as the dean of Soviet computing memory—claimed that "Glushkov's monumental efforts constantly ran into a wall of indifference, misunderstanding, and at times, animosity in the top echelons of the command-administrative system." According to Malinovsky, the Soviet higher-ups who never publicly criticized Glushkov were Prime Minister Kosygin and Defense Minister Dmitry Ustinov, although the proposal also elicited resistance from lower-level figures.[11] Nonetheless, after years of politicking on behalf of the OGAS Project, Glushkov convinced the CSA to reinstate the word *OGAS* into the 1976 report on its "Main Directions"—and breathed new life into the core idea of a Soviet industrialist network that united automated control systems across national economic branches.

A year later, in 1977, the state decided to declassify the OGAS Project, meaning that the OGAS was no longer a state secret. This decision reflected the project's declining strategic significance to the state as well as a shift in Glushkov's long-term campaigning. Before 1977, promoters accepted the ban on public discussion in part because it meant that the secret project was vital to the highest political echelons. But this secret classification also served its opponents in the state because public circulation and promotion of the OGAS could have curried public favor for what could prove to be a career-threatening reform. After the lifting of the top-secret clearance, however, this could change, and Glushkov successfully petitioned *Pravda* newspaper editors to begin a campaign to promote the network project with his article titled "The Matter of the Whole Country" in 1980 (although Malinovsky notes in the English translation of the dual-language *Store Eternally*, without clarification, that the published version of the title was actually "For the Whole State").[12] The article's publication in *Pravda* implies a mixed public relations victory because appearing in the nation's flagship newspaper meant that its editorial board, the Central Committee itself, had deemed the project to be worthy of public discussion and not one of its prized state secrets. (The conclusion to this chapter returns to the issue of public discussion.)

With a sizeable audience for the first time, the OGAS Project diversified quickly into a number of complex possibilities in the hands of leading academics such as Glushkov, V. A. Myasnikov, Yu. A. Mikheev, and others. Under their leadership and assignment to build a technical network that connected local factory control systems, a number of associated subprojects arose, including the ACPR (the automatic system of planning accounts, or *avtomatizirovannaya sistema planovyikh raschetov*), ASGS (the automatic system of state statistics, or *avtomatizirovannaya sistema gosstatistiki*), the ASUNTP (or automatic system of management of scientific-technical progress, or *avtomatizirovaanaya sistema upravlenia nauchno-tekhnicheskim progressom*), and the ASUMTS (automatic system of management of material-technical supply, or *automatizirovannaya sistema upravleniya material'no-tekhnicheskogo snabzheniya*).[13] The subsequent multiplication of associated ASU systems and subsystems in the late 1970s and early 1980s attests to two underlying trends—first, a general academic (public) interest in the OGAS Project across planning, statistics, science-technological revolution, and supply institutions; and second, a splintering or at least division of that overlapping interest into subsystems according to preexisting complex relations between branch, regional, and national economic planning interests.

The movement to "ASUify" the nation in the 1970s never met with considerable success. Given that the introduction of an ASU to a factory or enterprise costs on average about 800 thousand rubles (or roughly just over $1 million U.S. in the 1970s or over $4 million U.S. in 2016), ASUs were introduced slowly and steadily in the late Soviet Union. According to one account, as few as twenty-nine ASUs were introduced between 1971 and 1975, thirty-two between 1976 and 1980, and thirty-four between 1981 and 1985. Another report holds that from 1971 to 1975, the number of ASUs grew almost sevenfold, although, even if the OGAS Project were suddenly approved, they could not easily be unified.[14] Other accounts mention even higher numbers, including one that claims that between 1966 and 1984, approximately 6,900 ASUs of different configurations were established throughout the USSR.[15] This vast discrepancy underscores the point that whatever systems were developed under a sweeping state mandate to advance information technology throughout the country, they were done so without the benefit of any organized coordination from the state. The lack of coordination hurt the effectiveness of the OGAS Project. Official statistics determined that the computer technology that was in place fulfilled no more than a sixth of its projected capacity in the affairs of local economic management.[16] The introduction of more computing processing power in the form of third-generation computers adopted from abroad significantly altered these modest growth trends in the managing of the command economy. The effort to network local enterprises and factories was met with resistance from workers and managers. There was brooding factory floor–level discontent with the local factory computer control installments, which were the local nodes that someday might be connected to form the EGSVTs and OGAS.[17] The workers did not feel empowered by their access to the circuitry of the state's master plan because the master plan exercised managerial oversight over only local factories. As in early computing industries elsewhere, the simultaneous development of different core computer systems in different systems led to protracted interoperability problems and technical delays.

The state secrecy that characterized the OGAS Project from 1959 to 1977 facilitated a kind of boundless technocratic imagination about the possibilities of networked computing that was not tempered by the humbling revelations of practical experience. Although the early developmental period of the OGAS saw profound accomplishments (including the launching of satellites and astronauts in space, the harnessing of the atom, and the advance of problem-solving machines), when those technological innovations were applied to everyday routine operations and tasks (such as sending and receiving economic information across the command economy and developing automated programs for deciding what to do with that information), the experience of networked computing in the late Soviet period revealed just how modest the accomplishments of most science and technology are day to day. This widening breach between the grandiose intellectual possibility and the modest applied practicality was central to Glushkov's local-global approach to practical universals, but it had the unwanted effect of dampening public and institutional enthusiasm for the unmet expectations of the OGAS Project. Accusations flew among stalwart Communists, who blamed the internal divisiveness and infighting among the top levels of government on the interventions of skilled enemies, especially American capitalists. However overgenerous to enemies' prowess for subterfuge this may be, one sympathizes with their frustration while doubting the utility of such countercounter measures.

In their discursive move and countermove, the public debates about networks in the 1970s are not exceptional for the period and place. Just as Kitov, Lyapunov, and Sobolev had done in their initial article by claiming that anticyberneticist Soviet philosophers had fallen victim to the machinations of a subtle pro-American disinformation program, Glushkov occasionally partook in that classic cold war move of blaming the cunning enemy for one's internal problems. Glushkov, for example, once blamed an unnecessary political battle in the 1972 All-Union Conference on a "disinformation campaign skillfully organized by the American secret service, which was directed against the improvement of our economics." No matter how fueled by the fumes of international conspiracy, such claims appeared to work at home. Once, Glushkov reports, he was able to soften the blow of an internal attack on his local automated system of management (ASU) work by asking the Soviet scientific adviser in Washington, D.C., to issue a report on how the competitors to his proposed computer were becoming less popular in the United States. The report was read widely in the Politburo and had its intended effect, leaving Glushkov's project on the table and scuttling his competitor's.[18] The positive corollary abounds in practitioner memoirs, where a colleague compliments an associate by attributing retroactively visible similarities between friend and foe to the friend. For example, "as a thinker, V. Glushkov distinguished himself by the scale and the depth of his works," notes the president of the Ukrainian Academy of Sciences, Borys Paton (whom Glushkov served as vice president from 1962 until his death), and he continued that "he predicted many things that appeared in the Western information society much later."[19] While traveling abroad, Glushkov once declined a lucrative salary offer from IBM, which also stands as a badge of honor. In both, rivalries imprinted images of personal hopes and fears onto the faces of doppelganger foes.

Few among the technocratic optimists or the disappointed practitioners were prepared to make the more general observation that the OGAS
experience is not unusual in how its bold technocratic inventions and pronouncements were followed by plateaus of technological innovation that
swept through the long story of Soviet history of technology and science.
The real story about Soviet computing networks has far less to do with the
technology itself than with the institutional, political, economic, and social
networks that made up the knowledge base and innovation infrastructure
in a country and culture.

Bureaucratic Barriers

Glushkov was clear about the sources of the frustration to his life work: cunning enemies were not infiltrating his life work from outside the nation, but
cunning competitors from within were doing so. After the Central Committee's partial rejection of the OGAS Project in 1971, rumors circulated
that his local enemies were conspiring against him. In 1972, the pilot of a
plane that Glushkov was flying in had to make an emergency landing and
discovered that the fuel had been tampered with. It was rarely cloaks and
daggers for prominent Soviet mathematicians, however. The most common
obstacle was the pragmatic apathy that prevailed against his ideas for technological reform. In response to the proposal for an electronic office, for
example, a commentator expressed doubt: "if it takes a month and a half to
act on a letter to a Ministry, no automatic letter opener is going to change
anything."[20] In his memoirs, he calculated the malaise that characterized
his meetings with government officials with a characteristic precision:
"Unfortunately, my organizational efficiency coefficient ... did not exceed
four percent. What does that mean? It means that in order for a problem
to even be considered by the government, I had to speak with twenty-five
officials."[21] An inefficient bureaucracy was both the obstacle to as well as
the target of his technocratic reforms. In 1972, he illustrated this with an
eye-catching statistic: according to his estimates, at 1 million operations per
man with an adding machine, it would take "10 billion persons" "to solve
all of today's management problems." The same operational burden could
be handled by men and women at 25,000 to 30,000 Minsk-32 computers
(at 30,000 operations per second), and even that number would quickly
decrease as processing power continued to increase.[22]

Antibureaucratic sentiment is not uncommon among highly skilled
technical workers and even among other bureaucrats. In Glushkov (whom
Hoffman once described as "probably the most forceful Soviet advocate and

the bluntest critic of computerized communication in the USSR"), it took a particularly acute form. Again, recall that after the Central Committee heard that the ARPANET had gone online and effectively granted Glushkov a blank check, the secretary of the Central Committee asked Glushkov to "write down in detail what has to be done," said A. P. Kirilenko, "and we will create a commission." Glushkov's response is reminiscent of Baran before he withdrew his network project from consideration by the U.S. military. Both insisted that, whatever else happened, their network projects not be handed over to the available administrative entities."[23]

In over a dozen interviews that I conducted with scientists who were associated or familiar with the OGAS Project, they unanimously complained that bureaucratic infighting was the primary obstacle to their project. There was something dreadfully wrong with the bureaucratic administration of the national economy and the handling of the OGAS Project. But Glushkov's critique looks beyond the bureaucrats themselves. Given a Weberian understanding of bureaucrats as depoliticized professionals, many people who held positions in the command economy and state were not rational bureaucrats at all. They affected an "iron cage" of bureaucratic petrification when convenient and waged war with other local deities. The problem does not belong to all modern bureaucracy. Some bureaucracies do not result in this kind of incessant, internecine Hellenistic competition among the gods. Some administrations, including Soviet military ones, have successfully managed to fund, develop, and launch megaprojects, and most large-scale modern institutions are administered by functional bureaucracies.[24] Where lay the difference?

Glushkov believed that a successful bureaucratic system could be reformed and improved with information technological upgrades, but only with commensurate social and economic reforms. In other words, technical reforms to administrative systems without behavioral changes were condemned to fall into a kind of double-bind: no minister could manage a complex economy by paper, and yet no one with control over the papers at hand would agree to switch to a "paperless" virtual economy, which Glushkov had championed in the 1970s. Because the "chief content" of the computing revolution was no less than a cybernetic fusion of information-processing people and their machinery (in other words, "the appearance of an essentially new man-machine technology for processing information"), the success of any technical system reform would depend on social and organizational changes: "Since the circular flow of information is the basis for the functioning of any organization, [the information revolution] must be viewed primarily as a revolution in organization and management."[25]

For its main theorist, the OGAS Project could not meaningfully upgrade to the command economy technologically without also simultaneously reforming the organization and management of economic information. Unfortunately, this separation of reforms is what the Central Committee had repeatedly requested when it insisted that Glushkov begin with the technical computer network EGSVTs before developing the automated system of economy management that was central to the OGAS Project. He frequently warned that without commensurate structural and behavioral transformations of the economy, the introduction of information technologies would *slow* economic growth:[26]

The conservatism of the traditional technology for processing planning and management information leads to the intensification of "disorganized complexity" in the national economy and erects informational-organizational barriers to planned economic growth.... The problem, of course, is not just in the technology of organizational management. The economic mechanism plays a large (indeed a primary) role here.... However, it is important to emphasize that economic mechanisms (especially under socialist conditions) do not work by themselves in isolation from the organizational management system.[27]

In other words, perhaps the most direct cause for the failure of the OGAS to develop, according to Glushkov, was rooted in the same motivation that drove him to develop the OGAS in the first place—the observation that the effects of the modern information science and technological revolution cannot be separated from the social, economic, and organizational conditions that shape them. The lot of networked computing cannot be understood without the networks of institutions that first attempted to usher technological networks into being.

This approach identifies at least two complementary organizational barriers to the success of any attempt to systematic reform—centralized self-interest and the decentralized status quo in Soviet society. First, had networked computing been integrated into the fiber of Soviet society (which it was not), it would have compelled broad-based systematic social changes that could not have easily been isolated (as they usually were) to the military industries.[28] Because military interests maintained a strong self-interest in preserving military power (and not social or economic progress), these same organizations also actively resisted encouraging the development or sharing the benefits of networked computing technologies outside of narrow military applications. It was clearly in the military's self-interest to maintain centralized control over networked computing innovations.

Second, at the same time, the decentralized network of competing interests that governed nonmilitary administration also ensured that attempts to introduce networked computing into social and economic planning would break against well-organized centralized resistance from the military and broad-based and haphazard resistance from anyone in a position to benefit from the status quo. Because the OGAS threatened to reorganize the social and economic spheres of life into the kind of rational planned system that the command economy imagined itself to be in principle, it threatened the very practice of Soviet economic life: networked computing, in Hoffmann's analysis, "creates more choice and accountability and threatens firmly established formal and informal bases of power throughout the entrenched bureaucracies."[29] These two threads of analysis met in the friction between a formal centralized hierarchy and an informal, decentralized heterarchy. Both the military powers and the decentralized network proposed by Glushkov were clearly hierarchical in operation. But the actual workings of Soviet economic and social power were neither hierarchical nor market. They were heterarchical, dynamic, and continuously reconstituted in the interwoven political networks of social relations in the economic bureaucracy facilitated by the Communist Party.

When asked why he thought that the OGAS did not take, Glushkov responded with a comment that distinguishes military (space and atomic) programs from the civilian administration:

S. P. Korolev ("the chief designer" of the Soviet space program) and I. V. Kurchatov (the father of the Soviet atomic bomb) had a guardian on their side in the Politburo, and they could approach him and immediately resolve any question. Our trouble was that we had no one, and our questions were even more complicated because they involved politics and any mistake could have tragic consequences. For that reason, a connection with any of the members of the Politburo was that much more important.[30]

Aleksandr Stavchikov, historical secretary of the Central Economic-Mathematical Institute (CEMI), also commented on why the EGSVTs did not develop successfully. According to his unpublished notes and personal interviews, Stavchikov retroactively faults "the romanticism" of the institute for the "globality" of its early network designs, observing in hindsight how Glushkov, Fedorenko, and others agreed early in the 1960s that any attempt to plan the national economy in its entirety would have to be done at the national level: "Certainly, an attempt to plan the national economy of such a huge country on the foundation of one hugely proportioned economic-mathematical model," Stavchikov admits, "would be doomed to failure from the start."[31]

As for why the network was designed hierarchically, Stavchikov intimated that the cyberneticists had no better choice. The reasoning for the hierarchy—and not, say, a fully distributed design or even an unevenly decentralized or heterarchical model—was a matter of reading the writing already on the wall: Nemchinov and Fedorenko decided to "build the country's unified net hierarchically—just as the economy was planned in those days."[32] (The institutional histories of CEMI and the Institute of Cybernetics conspicuously leave out the names of Glushkov and Fedorenko, respectively, although evidence of the early alliance abounds in personal memoirs and interviews.) Justified by a grand cybernetic analogy between the formal design of the command economy and the formal design of the computer network, Stavchikov reasons, any other network design would have been politically unviable in a formally hierarchical command economy. The network visionaries had no choice but to design a computer network that matched a system that did not exist except, like the networks, on paper free from the informal competitive practices of administrative-economic reality.

Design logics can be compelling—too compelling at times. The cybernetic analog between hierarchical economy and network also fit the political values of the period. Fedorenko and Glushkov felt they had no other choice: they had to align their technical national architecture with the political system architecture. They also appear to have wanted to do so. All evidence suggests that these leading cyberneticist entrepreneurs were committed believers and practicing promoters of the official socialist rationales of the command economy, which also made them reformist critics of the irrational status quo. For these network entrepreneurs—Fedorenko, Glushkov, Kharkevich, and Kitov—the heterarchical competition at every administrative level was the signal problem that was in need of a sociotechnical fix. According to CEMI Secretary Stavchikov, "In this, [Nemchinov and Fedorenko] planned to use extant economic-mathematical methods, allowing [them] to guarantee mutual conformity, the very best interdependence of the numerous units of the hierarchy downward and horizontally—between the units of one level, as well as to develop new units."[33] In other words, according to Stavchikov, these economic cyberneticists decided to model the structure of the network after the structure of the socialist economy, in essence invoking the well-established trope of cybernetic thought that technical systems share common information structures with social systems—including mind-computer, body-machine, and society-media systems. The hierarchical form came as cybernetic analogic impulses such that the decentralized network proposal was designed, according to Stavchikov, to "guarantee mutual conformity" and "interdependence" between the formal Soviet economic hierarchy and its technical network.[34] The cybernetic instinct to design the OGAS after a nervous system for the national economic body and not the nation as a brain follows from this felt obligation to "mutual conformity" between the national economic and network hierarchies.

The choice of the national hierarchy as the basis for their network design was both the industry standard and necessary for those who were looking to streamline a command economy that was both hierarchical on paper and heterarchical in practice. The contradiction that was central to the breakdown of the Soviet economic-administrative system lies between the formal hierarchical design of that state and its own informal heterarchical networks of management as practiced by those who administered the state. The endgame of the OGAS Project was found not by strategic problem solvers who were seeking to solve or finish the game but by those who were seeking to extend perpetually their turn at the table of administrative power.

This is a distinct argument that is separate from the standard historical accounts of the collapse of the Soviet network projects and sociological accounts of the collapse of the Soviet Union itself. Most accounts posit that the basic problems were one of a rigid, top-down hierarchical state. Noted scholars since the 1990s have argued that the Soviet state and command economy were fundamentally incompatible with the emergent, flexible information networks.[35] I believe that I have shown why that is wrong and why it misses the greater problem. Instead of a fundamental incompatibility between vertical states and horizontal networks, the heterarchical ambiguities of Soviet administrative networks reveal too much, not too little, flexibility in its capacity to generate organizational dissonance crisscrossing and overlaying economic hierarchical structures with lateral conflicts of private interest. The Soviet state was too familiar with the unpredictable dynamism of competing informal networks (the same kinds of networks celebrated by Internet commentators in the 1990s) to be able to carry out systematic reform and infrastructure upgrade to bring the Soviet state into the current network information age.

The Red King's Book, or Botvinnik and the Soviet Case of Computer Chess

If war, in Carl von Clausewitz's famous phrase, is a continuation of politics by other means, then perhaps the most visible continuation of cold war politics by means of a game is chess (second to Go, the world's most popular war game). This classic thinking man's game is synecdoche for cold war confrontation, complete with two diametrically opposed rational strategists plotting the endgame of the other.[36] It is no surprise that the Soviet Union, which reigned as chess hegemony for most of its existence, took its strategic chess, computer, and long-term planning thinking seriously. Among those thinkers stands Mikhail Moiseevich Botvinnik, who, although not quite the brightest star in the constellations of Soviet grandmasters, is nonetheless remembered as the patriarch of Soviet chess for innovating and institutionalizing rigorous systems for gameplay. As this section explores, with the support of Glushkov and others, Botvinnik even programmed his own end game for cold war chess itself. His *Pioneer Project* stood as an attempt at computer chess programming that he felt would bring the Soviet Union one step closer to triumph in strategic political economic planning.

Raised in St. Petersburg, the son of a dental mechanic who had earned the right to move beyond the Pale of Settlement, and married to a ballerina (the other superior Soviet art of elegant maneuvers), Botvinnik (1911–1995) came to chess at the late age of thirteen and left the chess world a different place seventy years later.[37] In 1935, at age twenty-four, he became the first Soviet grandmaster, and by 1957, under his guidance, there were nineteen Soviet grandmasters, with roughly twenty new masters emerging every year. As a figure astride Soviet chess history, Botvinnik is remembered today for establishing "the Soviet school of chess"; for mentoring world-famous chess figures Anatoly Karpov, Vladimir Kramnik, and Gary Kasparov; and for promoting disciplined chess training (a cross of physical and mental exercise). He was also a theorist of long-term strategic planning who pioneered Soviet computer chess and chess schools based on his notational system for recording chess play. That notational system preceded what is now known in the chess world as "the Book." A fascinating character on his own right, Botvinnik figures here because he is an early Soviet network visionary. Like the other distinguished scientists and long-term strategists who were committed to the Soviet way of life, he proposed that the state use computers to optimize and resolve its long-term planning problems in economic and political spheres.

Botvinnik's combination of professional success and political notability was a rare distinction for advanced Soviet chess players, whose demanding careers as civilian celebrities rarely left time for anything else. He even was awarded a national medal of honor for his work as an engineer at the same time that he was establishing himself as a world chess grandmaster. In 1954, six years after defeating the reigning American to win the world championship, Botvinnik came as close to a public icon as the Soviet Union had then (the superstardom of cosmonaut Yuri Gagarin came later). Botvinnik received spontaneous standing ovations on entering movie theaters and was one among few other than members the Party elite who had a private

car and driver. Buoyed by such a reputation, he wrote a strong-willed letter to *Pravda* in 1954, the year after Stalin's death, detailing a long-term
strategy for world domination without having to go to nuclear war. His
suggestions involved calculated moves and countermoves through which
the socialist leaders would grant the masses of petty capitalist owners their
material wealth in exchange for their acceptance of the socialist revolution
without atomic combat.[38]

For such a brazen public stunt, the political secretariat rebuffed him and
threatened to throw him out of the Communist Party. In the 1960s, he
publicly repented and avowed his Communist credentials by publishing
Computers, Chess, and Long-Range Planning, which describes how the domination of the Soviet school of chess over the Americans was an expression
of superior long-range socialist planning. In 1968, having been influenced
by Claude Shannon's less well-known 1950 work on computer chess, Botvinnik published *An Algorithm for Chess*, which successfully demonstrated
how to algorithmically organize attacks against an opponent's position
from challenging tactical positions. Even though his algorithm excelled in
solving technically stressful positions, it also had the frustrating tendency
to overlook the simplest tactical moves.[39]

Backed by major cyberneticists and computer engineers including Viktor Glushkov and his colleagues—such as Bashir Rameev (who developed
the Ural computer series), Viacheslav Myasnikov, and Nikolai Krinitskiy
in the 1970s and 1980s—Botvinnik poured his energies into what he called
(drawing on the Stalinist vocabulary of his youth) the Pioneer Project, a
computer chess program that was designed to imitate how the brain of a
grandmaster works.[40] The OGAS Project algorithms were designed to ignore
the bulk of all computationally possible moves and instead to concentrate
on the most probable moves. Attempts were made to develop an algorithm
with a long-term intuitive "feel" for the board. The brute force approaches
(which calculate all possible moves in branching decision trees) eventually
won out with the arrival of faster computers in the 1970s, outpacing Botvinnik's selective but theoretically more sophisticated approach. Nonetheless,
his prodigies in training—Kasparov among them—remember their surprise
at hearing that Botvinnik was confident that his selective program would
one day consistently beat them all. Nevertheless, it—much like Glushkov's
attempt to develop intuitive macroprocessing and natural language programming that mimicked the neural processing and speech patterns—bore
fruit in other spheres of application. For example, Botvinnik, in his career
as an electrical engineer, reconfigured his Pioneer algorithm into planning
maintenance repair schedules for power stations across the Soviet Union.[41]

The Pioneer Project and the OGAS Project shared more than a common organizational framework and set of state-of-the-art computers. At core, they shared a commitment to organize the real-time management of scarce computational and economic resources. In the early 1980s, Botvinnik tried to salvage the national economy with another proposal that he sent to Party leaders. It contended that the Soviet economy should be regulated by a software program that, like his Pioneer chess algorithm, would take a generalizable approach to reasoned decision making. Botvinnik thus stands out as the last of major Soviet figures (with Kitov, Kovalev, Fedorenko, and Glushkov) to propose using computer software to salvage the command economy. Available records do not speak to his proposal's reception, except that it was rejected at the highest levels. By the mid-1980s, Gorbachev's reforms had already sufficiently introduced market elements into the formal command economy to render impossible any future systematic management of the economy. In the early 1990s, shaken by the collapse of the Soviet Union and years from death, Botvinnik reached out one last time with strategic advice for Yeltsin's government but to no avail.[42]

There is a truism in the history of science that science serves many specific social purposes but basic research need not begin with any single goal in mind. Biologists, for example, run test on fruit flies—or *Drosophila*—not because they are particularly devoted to improving the life of fruit flies; they do so because fruit flies are convenient test subjects that reproduce quickly and cheaply. Computer chess has been called "the drosophila of artificial intelligence" (Alexander Kronrod's phrase, popularized by American computer scientist John McCarthy) because it is thought to stand in as an affordable test case for larger strategic programming projects, which include both artificial intelligence as well as planning the Soviet command economy.[43] Kronrod, himself a distinguished Soviet mathematician and computer scientist, also collaborated with Kantorovich on the computer planning of the economy and with Botvinnik on the algorithm that defeated the Kotok-McCarthy American chess program in 1966 and 1967.[44] The unexpected joy of computer programming lay in finding new applications for old techniques, which in many ways was the same allure that fascinated general-purpose computer programmers since Turing. Although OGAS, EGSVTs, ESS, ESAU, and even Botvinnik's Pioneer Project "failed" on their own terms, they also should be remembered for their contributions to ongoing macrolevel experiments in rational planning, administration, and policy making in a world of global information networks.

As these cases suggest, the consequences of the current networked information revolution cannot be easily anticipated. The chess community has long concerned itself with the advancement of computer chess programs, which grew exponentially more sophisticated from 1970 through 1989, when Gary Fields was first defeated, and then again until Deep Blue's controversial victory over Kasparov in 1996. Since at least Wiener in 1964, critics have contended that superior computer chess programs were inevitable but would diminish the value of chess as a human activity.[45] The situation led early computer chess critics to bemoan their state with the defeatism of the final scene in the 1980s film *War Games*. After all, if the heirs of Botvinnik's Pioneer program dominate the best players today, the human will to be the best has already been undermined. What is the point of playing, the chess enthusiasts worried, when everyone loses every game?

Such handwringing by chess purists against the artificial intelligence community has since been sidelined—and by neither the triumph of technology over humanity nor the triumph of humanity over technology. Chess as a human pastime has not dwindled in the face of virtually indomitable computer programs. Instead, networked computers have sped the spread and growth of the global chess community. The number of online human-to-human and human-to-computer chess has exploded since Kasparov's defeat for unrelated and seemingly mundane reasons. No longer encumbered with the burden of serving as a shadow stage for cold war intrigue, long-distance chess in real time over computer networks is now an everyday reality.

Botvinnik's influence on networked computers and chess continues to surprise. It is not Botvinnik's sophisticated computer algorithm but his foundationally basic notational system that has had the most lasting effect on the now globally networked game of chess. Thanks to well-codified chess notation systems that were popularized by Botvinnik, computer record-keeping capacities have allowed millions of games of top-level chess to be catalogued into a database known as "the Book." Recently, for example, a German company named ChessBase has been scrutinized for its widely used database of chess moves that organizes prior games, new move opportunities, and errors in human play, which effectively reduces chess games to enormous decision trees of known and unknown pathways of game progression. Critics have accused its founder, Frederic Friedel, of having "ruined chess" because few games now occur that include new combinations of moves that are not found in "the Book."[46] The result is a new cold war tension of human players against the book, in which top chess players and their opponents know that, given almost any chess board arrangement, the best game they can play is played out in "the Book." Botvinnik's secret library of index cards that recorded global grandmaster games, saved exclusively for study by students at his Soviet school, did not migrate online, but it modeled what has become the global networked norm. Top players worldwide now memorize tens of thousands of recorded games and positions. All chess competitors now play aware of the networked heir of Botvinnik's book and the humbling fact that most chess sequences have already been played before. The introduction of networked computing is driving a curious situation (however common when new media become mainstream) in which Botvinnik's dream has now been achieved (for example, since 2005, the best software programs routinely trounce the best humans at chess) without appearing the affront to humanity its critics predicted it would be. In fact, global communication networks have made correspondence chess (with humans and computers alike) more popular. Perhaps the enduring attraction to strategic pastimes reveals, with a gesture to Walter Ong, that there may be nothing more human than artifice. (Consider the complex rules and recipes behind baseball and apple pie.)[47]

Computer cold war chess offers a view of the historical preoccupation with global and long-term planning strategies from Liebniz to modern-day generals.[48] The reformist efforts of Kantorovich, Glushkov, Fedorenko, Kovalev, Kharkevich, Botvinnik, and many others are not exceptional. Rather, the introduction of the digital network in socialist cybernetic planners and the sharing of "the Book" in chess underscored something that was at once peculiar yet normal. Networks make knowledge generalizable or at least generally shareable and remixable—whether a dataset shared by network or a playbook shared within a chess school. The consequence of that record, after it was repurposed from the secret index files of Soviet libraries into open-access public repositories, in turn is purported to do nothing less than remake the chess world. Such a grandiose sentiment outlines the strong intellectual affinity between Soviet cybernetwork visionaries and the modern preoccupations with network-enabled public recordkeeping and its automated extension, surveillance.

Simultaneously, the experience of Soviet computer chess also underscores the critical fact that, although military and civilian projects in the Soviet Union suffered from being strongly separated, the cold war culture—especially cybernetic tools, game theoretic strategic thinking, and the computational management of limited resources—has spread the influence of military and strategic thinking far and wide into everyday matters of politics and economics. In chess as in planning, the separation of military and civilian administration offers no guarantee of the same in modern society.

How Hidden Networks Unravel Cybernetworks

This chapter has introduced and advanced an argument based on the informal character of the Soviet system outside of the centralized military command. The dynamic vitality of the system—unregulated competition with unpredictable promotions and demotions—did not always benefit the well-positioned and talented network entrepreneurs and system reformers, as a number of case studies have shown. As Kitov's Red Book show trial demonstrates, military superiors were free to punish their own best and brightest for suggesting that networked computing capacities should be shared beyond narrow military applications. His superiors formally accused him not of displaying generosity toward civilian concerns but rather of going outside formal military communication channels, which underscores the depth of the structural military-civilian divide behind the Soviet networking story. The early partnership between Glushkov's Institute of Cybernetics and Fedorenko's Central Economic-Mathematical Institute illustrates something of their double-edged situation. Glushkov and Fedorenko faced opposition from the centralized military command of the Ministry of Defense, which denied them access to military networks and the institutional knowledge base that supported those military networks. At the same time, they also faced a more subtle institutional obstacle to OGAS that came from the civilian economic sector. The unpredictable currents and institutional drift of the state bureaucracy, including a flush of untethered funding, pulled their young, growing, and capable research staffs in divergent directions— including a focus on macroeconomic reform under Glushkov in Kiev and a focus on microeconomic reform under Fedorenko in Moscow.

Near the end, Glushkov reflected on the sources of the obstacles that his team faced when they were developing the OGAS, the EGSVTs, and national economic reform. His sense of disappointment with his own generation was particularly acute, and his last book that was published while he was still alive targeted schoolchildren as its audience—*What Is the OGAS?*[49] In 1983, while on his death bed and suffering from an apparent tumor of the medulla, Glushkov proclaimed that the OGAS was his "greatest life work," after over twenty years of dedicated effort and a long list of significant accomplishments in other major scientific fields.

During the last nine days of his life, while constrained to his hospital bed in the Kremlin, Glushkov insisted on working—just as he had done back in his hospital bed in 1962. During those final days he dictated his life memories to his daughter Olga and received as a guest the deputy of

Dmitry Ustinov, one of his staunchest supporters in the military. Ustinov, Glushkov reflected, had managed to do in his military career what Glushkov could not do in the civilian sector—rise from the chair of the Supreme Council of the National Economy under Khrushchev to wield power to reform the Ministry of Defense as its minister and marshal of the Soviet Union. Ustinov's deputy listened to the dying man's account of the "long ordeal" of his constant skirmishes with state bureaucracy before asking what the minister of defense could do to help. Glushkov, wrapped in the tubes of respiratory support, sat up and growled a memorable deathbed witness to military might and its remove from civilian concerns, "Let him send a tank!" Before an excessive growth in his own nervous system could bring down this champion theorist of the Soviet economic nervous system, Glushkov tried to comfort his grieving wife, Valentina. In his hospital bed in the Kremlin, he turned to her and spoke about the possibilities of immortality: "Be at ease," he said. "One day the light from our Earth will pass by constellations, and on each constellation we will appear young again. Thus we will be together forever in the eternities!"[50]

After Glushkov died on January 28, 1982, the OGAS vision continued to radiate outward and did not immediately fade from state discussions, social networks, and print media. Anatoly Kitov attempted to reanimate the proposal by writing directly to General Secretary Gorbachev in October 1985. Kitov, then the chair of the Department of Information Technology at the Plekhanov Moscow Institute for the National Economy (part of the Russian Academy of Sciences), recounted the history of the OGAS Project— the scattered development of unconnected ASUs in the 1960s and 1970s, Kitov's repeated appeals to the state for support, the subsequent disappointment with the spread of ASUs and the potential for networking them, the lack of state coordination over technical as well as administrative matters (especially the cooperation problem among separate ministries), and the fact that "we do not have modern reliable personal computers." "I think that this report constitutes an objective analysis of the last thirty years of developing information technology," Kitov concluded in the letter: "may it bring specific benefit and capacity for further decisive action."[51]

This time, Kitov's 1985 letter was not intercepted, although his reclamation of the OGAS situation came to the same effect as his Red Book letter did almost thirty years earlier: nothing would be done. The way that he was told this, however, reveals a crucial look into the inner workings of the administrative state. On November 11, 1985, Kitov received a phone call from Yu. N. Samokhin, a representative from the economic division of the Central Committee that had reviewed his letter to General Secretary

Gorbachev. Kitov's notes record that he was told two things: first, he was to be thanked for his contributions, and second, "not everything in the letter is supported by the economic division." The Politburo and the Central Committee, he was told, "had other functions, not those of the automatic management of the command economy." The Politburo was already supporting the creation of a state committee of information technology, and at the moment, that, not the economy, was the state's priority. Kitov, at the end of the telephone conversation, asked to receive the reply in writing and was told that the Central Committee did not provide written replies.[52] The likely reason for not offering a reply in writing was that the Central Committee did not want to proliferate in writing its own contradictions—in this case, that the economic division of the governing body of the Soviet state does not concern itself with the automatic management of the economy. No doubt Kitov felt that this reply was begging the question: that, it seemed, had been precisely the problem all along.

Such telephone revelations, however, did not keep the state, one year later in 1986, from pronouncing with the force of law that the economy actually would pursue the following Glushkovian demands over the coming five years (in the twelfth five-year plan): it would double the level of automation, organize the mass production of personal computers, increase the installation of computers by 100 to 130 percent, build computer centers for collective use, create integrated information banks, and significantly increase research in information theory, cybernetics, microelectronics, and radio physics.

The passage of time has allowed some reflection on the sources of these challenges. In 1999, Fedorenko contemplated the stubborn fact that decades of CEMI efforts to develop macrolevel economic models had born very little fruit in part because "the problem was too multidimensional and multifactorial." But "the very hardest," Fedorenko admitted without clarification, "was the 'human factor.'"[53] Three decades earlier, the problem was effectively the same. In 1968, Kitov summarized his own frustrations in a personal letter to Lyapunov, not so much as a problem of human personalities or specific personnel but as a problem of cultural resistance to reform in the institutions:

The top leadership realizes the importance of [the introduction of computers into the national economy] but takes no effective measures in support of such work, while responsible officials from the ministries and other government agencies ... display no interest in the automation of management for the optimization of planning. The problem is apparently rooted not in their personalities, but in their positions and in the overall traditions, which change very slowly.[54]

Here Kitov faults not the top leadership of the country but the institutional logic of administrators in the middle levels of ministries. Kitov's complaint holds that middle managers habitually reaffirmed the status quo (as is also the case in many large organizations) and that the dynamics within such conservative institutions were paradoxically nonsystematic. As countless Soviet officials have observed: "Having different ministries is like having different governments," and the battle between civilian ministries often flares into nothing less than internecine civil war.[55]

The paralyzing competition between these dynamic, unregulated ministries constrained the possibilities of both systematic institutional growth and purposeful reform. Interministry cost-sharing and cooperation rarely happened. Whenever high-ranking Soviet administrators wanted to promote a major project (such as a national network), the primary avenue for action available to them, as Garbuzov's counterproposal anecdotally indicates, was to create an entirely new institute within preexisting administrative silos. Thus, the attempt to create a supervisory institute for a particular sphere of responsibility (such as finance, statistics, or the OGAS) created intractable points of competition between those institutes. Instead of easing the conflict among administrative standards, every new umbrella institution introduced a new competitor and exacerbated the power skirmishes. The attempt to create a hierarchical bureaucracy to resolve conflicts of administrative interest often generated more, not fewer, opportunities for infighting among neighboring bureaucracies. So the chasm between military and civilian administrations was perhaps not entirely insurmountable: while the military kept the country ready for war with the enemy, the civilian bureaucracy was already at war with itself. Unable to receive the same preferential state treatment as the military, the Soviet economic bureaucracy militarized itself against itself.

The uneven economy of those administrative silos often pivoted around surprisingly few well-placed administrators and veto points. Consider, for example, that the tenures of the chair of the Council on Cybernetics Aksel' .Berg and the mathematician Mstislav Keldysh, president of the Academy of Sciences of the Soviet Union (1961–1975), coincided with the rapid growth of Soviet cybernetic academic preoccupation in the 1950s and 1960s. Together, Keldysh and Berg personally facilitated the creation of all four main institutes featured previously, including the Computation Center 1 that Kitov directed following his optimistic report about the future of computing technology in 1953, the Institute of Cybernetics in Kiev under the directorship of Viktor Glushkov (1962), the Central Economic-Mathematical Institute in Moscow under the directorship of Nikolai Fedorenko (1963), and the Institute for Telecommunications in Moscow under the directorship of Aleksandr Kharkevich (1963).[56] Given the immense reach and corresponding tangle that is Soviet cybernetics, a disproportionate amount of its administrative growth took place at one or two degrees removed from these men's signatures and oversight. In the institutional growth period of Soviet cybernetics from 1953 to 1964, the roles played by supporting administrators like Keldysh and Berg helped extend the argument that Soviet institutions often experienced periodic spikes of exceptional growth followed by long periods of underdevelopment.[57] Perhaps the most striking record of the explosive state imagination for computing technology, at least as of 1963, that propelled Soviet cybernetics, CEMI, the Institute of Cybernetics, and the associated OGAS Project into the mainstream of Soviet political system is in the recently uncovered Party resolution published on May 21, 1963. This resolution, issued by both the Central Committee and the Council of Ministries, declared that the Soviet state would advance nearly twenty nationwide new or transformed tasks and institutions involving computing technologies, including the reform by computer network of the command economy.

Even so, the economic bureaucracy proved less resolute about embracing such sweeping technological reforms. As a house divided, the bureaucracy was unpopular with practically everyone, including the underserved public, scientists like Glushkov and Liberman, and politicians (such as Mikahil Gorbachev) who publicly ran against the bureaucracy not in earnest hope of reforming it but to ensure their own political popularity with the public.[58] In the main, the old guard of orthodox planners who administered the system benefited from it, although it would be a stretch to say that they approved of how it functioned. Crisscrossing structures, personal favors, and impartial administrative reforms plagued the hierarchy that held together the national, regional, and local planning committees. This ensured that, despite the state's approval of a single national plan, there were as many contested plans as there were administrators of the single plan. (No plan can plan away its own private interests.)

By the time that Gorbachev's perestroika and glasnost policies were introduced between 1985 and 1989, the national economy could no longer mobilize around mathematical economic reforms. Official statistics hold that between 1986 and 1988 the economy grew by 2.8 percent and in 1989 by 2.4 percent, although in practice real economic progress, like official economic statistics, was not meaningful. By 1985, after perestroika's decentralizing reforms (according to M. S. Shkabardni, one of Glushkov's colleagues), the idea of an economy that was rationally decentralized by OGAS "interested no one. Everyone had forgotten about it. No one even thought about it."[59] The Party leadership had many more pressing worries to consider, including the high capital investment that would be required if the economy were to be reformed by networked computers amid a rising stream of affordable personal computers from the West. The OGAS had long appeared a prohibitively expensive "hero project," but now even the more modest EGSVTs network could be built out by individual citizens who were working on Western computers. In a sense, that is what happened: large state network projects were abandoned, and in the late 1980s, a few Soviet citizens joined in purchasing and connecting personal computers to globalizing communication networks. By 1989, private Soviet citizens began logging onto early Internet chatrooms, by which time the OGAS Project, like the Soviet state, was slipping into history.

In summary, the OGAS Project was shipwrecked on the capricious unregulated conflicts of self-interest that occupied the civilian knowledge base (including but not limited to the economic bureaucracy) of the Soviet system. It fell prey to the conflicts of interest that it sought to set aside with automated networks. The sources of those conflicts arose from the yawning disconnect between the formal plan for the civilian sector, which was clearly hierarchical, and the massive gray economy of informal exchange and personal favors. Each layer of the command economy—the national, regional, and factory planners and managers—benefited from a slack and informal freedom that allowed them to solve problems outside of the plan's commands. By rationalizing, making explicit, and automating those resources, Glushkov's vision directly opposed the informal economy of mutual favors that oiled the corroded gears of Soviet production. In the end, the OGAS Project fell short because, by committing to rationalize and reform the heterarchical mess that was the command economy in practice, it promised to encourage the rational resolution of informal conflicts of interest—which worked against the instinct to preserve the personal power of almost every actor that it sought to network.

Conclusion

The portrait of the final chapter of the OGAS Project that is presented here fills out and begins to complicate the conceit with which this book began—that global computer networks arose from collaborative capitalists, not competing socialists (or in light of the OGAS Project, not from the unregulated conflicts of self-interested socialist institutions). This surely is no plain victory for any political order, nor is it only a plea for virtuously regulated market-state interactions. Self-interest has been a recognized engine of human behavior since at least the ancient Greeks and found in any economic order. (I understand self-interest here to be an ambiguous quality that is more basic than any particular economic order. It can range from a virtue as distinct from selfishness as satisfaction is distinct from hedonism to, as the Gautama Buddha taught, a signal vice of enduring dissatisfaction in life.)[60] The Soviet socialism that the Project sought to reform never worked as it planned in part because of the economic administration's mismanagement of its own conflicting internal egotisms and mutinous ministers. Its political economic tragedy lies in the flooding of the gray economy with the informal self-interests that the planned interests of the command economy—especially a technologically rationalized one—could never accommodate. It was not the absence but the presence of vibrant unregulated markets of conflicting forces driven by self-interested administrators that kept the Soviets from networking their nation and command economy. In another sense, the Soviet networked command economy fell apart not because it resisted the superior practices of competitive free markets but because it was consumed by the unregulated conflicts among institutional and individual self-interests—including the institutional rivalries that sprung up between Glushkov and Fedorenko's competing efforts to network and model the economy, the ministry mutiny over funding between the Central Statistical Administration and the Ministry of Finance over the network plans, and the adhocracy of the Politburo.

There is a problem, however—not with the history but with the ends of such critical analysis that inverses the role of regulated capitalist states and unregulated socialist economies. In so doing, it recapitulates the liberal economic coordinates for imagining the state as the site for public interests and the market as the site for private interests. The conclusion to this book outlines several reasons that such an analysis, although tempting, cannot hold on its own. Before that conclusion, let us summarize a few larger points that previous chapters have built toward.

First, the Soviet economic system did not work—except for when it did, which was mostly for highly centralized militarized projects. It is reasonable to presume, as social scientists and cyberneticists alike have been doing, that the Soviet formation of socialism cannot be separated from the economic and political woes that arose due to underlying structural contradictions. For the most part, those contradictions have been framed in terms of private (usually market) interests that were in competition with public (usually state) interests. Given this framework, the history outlined above may prompt defenders of private (market) interests to offer reminders about how, in the Soviet Union, private and public sectors managed at

best an "uneasy coexistence" or about how four decades after collectivization, private (market) plots that comprised 3 percent of Soviet agricultural
lands managed to produce nearly 30 percent of the gross value of Soviet
agriculture.[61] To argue that market solutions would work better does not
begin to describe or distinguish what I believe the OGAS Project history
reveals to be the depth and range of private interests at work in human existence. It is just as easy, in what prominent economist Igor Birman endorsed
as Soviet "anecdotal economics," to list examples of how the same kind of
private self-interest that put bread on the table of starving peasants also cor-
rupted socioeconomic life elsewhere. Anecdotes from everyday economic
life relate that 80 to 85 percent of gasoline, according to some estimates,
turned up on the black market;[62] construction workers built new apartment
buildings to state specifications but refused to connect the toilets to the
sewage systems until *vzyatki* and *podkupki* (bribes) were paid; maternity
nurses extorted 200 ruble notes from birthing mothers before using a sterile
needle and anesthetic; grieving families had to pay 2,000 rubles to bury
their mother, despite the guaranteed "free" state funeral and burial. The
fact that most numbers were anecdotal suggest how actively corrupt Soviet
economic life already was.

Self-interested corruption is so much a feature, not a bug, of Soviet economic life that it cannot be the result of market absence or state failure
alone. Anecdotes of administrative cunning (not incompetence) abound.
In Grossman's phrase, "the Four B's: barter, black market, *blat*, and bribe"
summarize the economic engine of Soviet self-interest run amok.[63] An
entire biscuit factory once went underground in Georgia, producing four
times its planned quota through hidden informants, bribery, and social
screens;[64] a seat on the trade committee in Moscow sold for 50,000 rubles
in 1990 (and current prices for other positions can be found online today);
and Central Committee members filled foreign bank accounts by extracting
bribes from officials in the trade ministries.[65] The Soviet joke puts it well:
Brezhnev is showing his mother how well he's done, and he shows her his
suite in the Kremlin, his dacha in the country, his Black Sea villa, and his
Zil limousine. "All very nice, dear," she says. "But what will you do if the
Bolsheviks come back?"[66]

These anecdotes constitute what we might call revolts in miniature. They
are an expression of private unrest—of local resistance to a society whose
public institutions did not have to serve the public. A liberal economic
analysis to these problems might describe the informal networks of competing private interest as variously productive or rent-seeking, depending
on whether the activity at hand created or depleted economic resources.

Varied critics of the Soviet economy have interpreted the collapse of the public interests of the state and the private interests of the market into the command economy to be a hallowing out of means for Soviet citizens to seek their own self-interest through formal mechanisms.[67] Consequently, informal means, whether creating islands of penny capitalism or engaging in systematic corruption, are all that is available.[68] The liberal economic critique accuses the public state of systematically smothering and driving underground private self-interest. Applied to the OGAS case, the standard critique follows: the OGAS Project could not hope to reform the command economy because its very purpose ran counter to the interests of those who held hostage that economy in need of reform.

The argument advanced in the conclusion to this book seeks to go one step further. It seeks to rearrange our thinking about cold war networked culture by twisting the standard liberal economic distinction between public states and private markets to feature a classical distinction between public *polis* (community) and private *oikos* (household). Instead of seeking to place blame on either the state for publicly stifling private self-interest or the individual bureaucrats for seeking to protect their professional self-interest by opposing reform projects, I suggest that the OGAS history reveals a third approach to social reform. The OGAS Project sought technocratic reform that is both public in its relationship to the market and private in relationship (or privy) to the state. It does not matter whether one faults the public state or the private market elements in the command economy that the OGAS Project tried to reform because they belong to the same classical category of private interest. Both state and market actors, collapsed into the Soviet command economy, sought their own private self-interests with certain consequences for how social and technological networks shape one another.

As the Soviet network stories show us, cold war economic orders prove more compatible in practice than in liberal economic theory, if for nothing other than their shared liability to collapse without careful regulation. Neither American-style capitalism nor Soviet-style socialism should be considered a sufficient philosophical banner for making our way into a networked world. If there is a shared baseline, it must be found in the agreement to regulate and restrain self-interest that is common to the visions of both Smith and Marx. The social necessity of restraining self-interested competition unites, not divides, the modern legacy of cold war socialism and capitalism. The following conclusion explores a few consequences for reintroducing a search for the role of *public* interests, in a classical sense of the term, in Soviet as well as contemporary network worlds.

Conclusion

InterNyet

This is the story—told for the first time in any language in book form—of a particular path not taken into the modern network age. Soviet scientists— led by Viktor Glushkov and his OGAS team between 1959 and 1989—could have developed a computer network project that brought about significant political, economic, and social changes. Had they done so, the current global network culture could have looked very different. Why did these network entrepreneurs not succeed? On what factors did the tragic twists of the tale we might dub the Soviet "InterNyet" hang?[1]

Faced with a struggling command economy, attempts to revitalize Soviet cybernetics, and a search for societal reforms after Stalin's bloody governance, Soviet researchers proposed as early as 1956 that computers should be used to control economic decision making. No one proposed that these computers be connected, however, until Anatoly Kitov, the military scientist who had "discovered" cybernetics in 1952, proposed in 1959 that civilian economists use existing military networks to solve economic problems, for which suggestion he was promptly dismissed from the army. At the same time as Kitov was making short-lived network proposals, Gluskov teamed with him and others to propose in 1962 a complex three-tiered hierarchical computer network that would transfer economic information along as many as, in its most ambitious proposal, twenty thousand local computer centers, several hundred regional centers, and one central computer center in Moscow. Over the years, this prohibitively expensive proposal was scaled down (and back up) to match the political climate. Nevertheless, the goal of this interactive, remote-access network remained the same—to reduce the coordination problems that had long beset the command economy. On and off over the next twenty years, Glushkov's OGAS team met resistance from at least five groups: (1) the military wanted nothing to do with civilian affairs, especially when that meant fixing the command economy that already fed its coffers; (2) the economic ministries (particularly the Central Statistical Administration and the Ministry of Finance) wanted the OGAS Project under their control and fought to the point of mutiny to keep competing ministries from controlling it; (3) the bureaucrats administering the plan feared that the network would put them out of a job; (4) factory managers and factory workers worried that the network would pull them out of the informal gray economy; and (5) liberal economists fretted that the network would prevent the market reforms that they sought to introduce. Instead of a national network, dozens and then hundreds of local computer centers—or automated management systems (ASUs) were built in the late 1960s and 1970s, although they were never connected. Thus the dream of networking Soviet socialism into a brighter communist future did not come to pass. This conclusion remarks on why this never happened and then hazards a few concluding comments and pronouncements.

There are many reasons why there were no such Soviet networks. But first is a reason to care about this story. Soviet network history invites us to think about the historical conditions of national computer networks without the assumptions behind the rise of current global digital networks. In other words, the OGAS story is a test case in how network projects could have developed in societies that were not preoccupied with markets, democracies, and personal liberties. Network projects without political and economic liberal values are not condemned from the start. Instead, after these cases are examined on their own terms, they can help control for, challenge, and rethink the conditions of possibility that are assumed to govern digital global networks. The Soviet network projects did not fail because they did not possess the engines of particular Western political or technological values. They broke down for their own reasons.

And these reasons were not the popular Western misconceptions. The standard criticism of Soviet technological backwardness (technological "behindness" would be more accurate) cannot describe on its own what prevented Soviet civilian networks from developing because the Soviet military possessed functioning long-distance computer networks since the mid-1950s and local area networks were linking ASUs since the mid-1960s. The technical know-how was in place. Nor can it be that computer networks are somehow inimical to closed cultures because computer networks have been serving military, authoritarian, and cybersecurity cultures for decades. That said, the history of Soviet technology overflows with technical problems—such as a lack of interoperable hardware or software for ASUs. Almost never, however, does the root explanation for Soviet technological problems lie in sheer technical incompetence. This advanced superpower state provided strong support for science.[2] The root problems with technology are anything but technological.

Beyond the Binary: Arendt and OGAS

Why was there no Soviet Internet? This book holds that leading Soviet scientists and their supporters—especially the OGAS team lead by Viktor Glushkov—tried repeatedly but could not network their nation with computers due to entrenched bureaucratic corruption and conflicts of interest at the heart of the system they sought to reform. McCulloch gives us a fresh term: *heterarchies* of conflicting private interests stalemated virtuous attempts to reform the hierarchical economic bureaucracy. If the Internet is not a thing but an agreement, as the phrase goes, perhaps the Soviet Internet is not a thing but a disagreement. (There is often more to learn from the latter than the former.)

This thesis, which expands on the standard interpretation, can be taken further. The history of the OGAS Project is akin to the history of a miscarried effort to perform an IT upgrade for the corrupt corporation that was the USSR itself. USSR, Inc., in other words, functioned as the world's largest corporation, and its private interests were internal market capture, the avoidance of the transaction costs of the capitalist market, and the concentration of power to itself. The political need for the OGAS Project appears to represent the grander inability of the hierarchical state structure of socialist politics since Marx to build and sustain innovation and reform in the age of industrial and information capitalism that the Soviet Union straddled. The network reform effort did not take into account its own effects on the formal command economy because the OGAS Project ran against the private interests of those who governed within an informal mixed economy. The perpetual conflict of self-interests that were internal to the Soviet system helps describe the continuous institutional tumult, frequent and ineffectual reforms, and currency of informal influence that underwrote the supposedly staid Soviet bureaucracy. The root problem here appears to be not the cold war binary between international economic systems but the binary that was internal to the Soviet economic system. Hidden, informal, and often vicious administrative networks prevented public, formal, and potentially virtuous computer networks from taking the Soviet Union online.

This view that the Soviet Union can be understood as a corrupt corporation also has its limits. In theory, it reads Soviet network history as it would read a Western state. In practice, it risks using the liberal economic values of market, state regulation, and individual interests to criticize socialist values of state-managed economies and collectivized interests—in effect, rehearsing the very political economic divide that it seeks to revise. This view falls short of explaining the motives and behaviors of other relevant actors. Although it depicts the perspectives of both the internal reformer (especially the scientists and administrative supporters of the OGAS Project) and the external critic, the interpretation does not describe why the militarized state, economic bureaucracy, and citizen workforce actively opposed ideologically faithful network projects. Why were their private interests in play at all, how can that question be described without rehearsing the exhausted cold war showdown between markets and states, and how might our answer to that question help focus critical attention on the contemporary scene?

Let us tweak our terms to state the situation more clearly. The OGAS Project could not achieve its end goal of reforming the Soviet economy because the hulking households of private power—the military, the corporation, and the state—compelled it into serving their private economic, not public political, interests. Consider the language of Hannah Arendt's *The Human Condition*—a landmark work of political theory that introduces its disenchantment with normative liberal values with a discussion of *Sputnik* and the nuclear age, the two ingredients that, once combined, could spell instantaneous planetary annihilation. For Arendt, the distinction between the public and the private is not the liberal economic opposition of the public state and the private market[3] but a classical (Aristotelian) distinction between the public as an expression of the *polis* (where actors gather "to speak and act together") and the private as an expression of the *oikos* (Greek for *household* and the root of the word *economy*) (where actors inhabit a domain of animal necessity and are compelled to pursue their own interests for their survival). For our purposes here, the *oikos* includes several institutional actors that usually are thought to be "public" yet that seek private interests for their own survival: the Soviet military men, with state backing, wielded the threat of nuclear destruction and personalized violence on the modern world; the Party leaders pursued their own interests independent of the people; the economic bureaucrats secured their own welfare apart from the welfare of the economy; and the citizen workers tried make ends meet in their private lives. The *oikos*, or the domain of the private, saturated the larger OGAS situation, and the history of modern networks, including but not limited to Soviet attempts, can be reread as a tale of private forces run amok.

These terms reframe our portrait of the challenges that were faced by Soviet network projects. The problem was not that the state failed to regulate private interests but that (according to Arendt) Marx put on a pedestal the speechless laborer (*animal laborans*), not the enlightened actor. The socialist state served and scaled up the most private and basic of human needs but no more. For Arendt, the equality of workers is tautological in the sense it equates people on the basis of animal need, and the equality of citizens should be sought by leveling unequal humans to create a better common world. She also targeted elsewhere the teleological violence rendered by Hegelian historical ideologies, such as Marxist-Leninist dialectical materialism: any state convinced of its own historical path is sure to bring ruin to itself and others.[4] In fact her critique of what she calls the rise of the social cannot be reduced to the ruinous rise of socialism (whether Soviet, German national, or other form) because her terms describe a range of modern advanced states that have led the ongoing global scientific-technological revolution.

For the purposes of this book, the rise of state and market as parts of a larger private household suggests the purpose of the command economy in both theory and practice. In theory, it collapses private economic interests into matters of state, and in practice, the state bureaucracy collapses into the institutional turmoil of private actors. We can also see that Communist Party leaders worked feverishly to secure their own power above all other concerns and that the military shielded the Party, spied on foe and friend alike, cannibalized resources, and separated itself from the national economy. The name of the Komitet Gosudarstvennoi Bezopasnosti (KGB) (Committee for State Safety) is similar to the name of the Committee of Public Safety during the Reign of Terror in the French Revolution, except that the Soviet version openly protected the state, not the public. The minister of defense made a related point in 1965 when he rejected any collaboration with the nascent OGAS Project. He identified the "healthy body" that his military served not as the public but as "the government mechanism," calling the economic welfare of the nation "a scab."[5] In view of the divisions in the Soviet *oikos*, this odd metaphor that the military was the mind of the state body (not that the state was the mind of the economic body) appears suddenly sensible.

Arendt's concerns about the escalation of private interests over public ones also explain why the OGAS story was not a people's history and why Glushkov addressed his last book to children, admitting that the workers were not prepared for the OGAS. Soviet citizens lacked mechanisms for mobilizing political will at scales larger than the dinner table, dacha, and press editorial, so they had few chances to observe a public hearing of the OGAS Project and far fewer chances to live a public life (or *vita activa*, as Arendt fancied it). By rotating the private-public distinction from one of market and state (and the state-market contradictions of the cold war economic order) to one of survival and political action, our vocabulary maps onto more private divisions in the Soviet household. The problems besetting the modern human world are far bigger than can be understood from any particular pole of the cold war and may even be shared between the two, as Arendt observed while she was in the middle of it.

Her argument comes with limitations. Like most political theory and commentary, it offers no concrete proposals for reforming the current situation. It idealizes a *polis* of ancient Greece that did not exist. It also gave no credit to the meaningful, vibrant, and even mischievous private lives that Soviet citizens experienced in the workplace, such as the Cybertonia case study (although Arendt notes that social gatherings that aggregate private interests can be charming but never glorious, a fitting summary of almost all virtual worlds and social media ever since). Moreover, her framing of the rise of the social cannot be used to describe the asymmetric inequalities of capitalism and social wealth because of the limitations of her founding image of the *oikos* as rooted in the private household. That image of the *oikos* would need to be subjected to a feminist philosophical critique of the power inequalities that are buried in the history of the household and domesticity—a critique that falls outside the scope of this book.[6]

To admit disillusionment with the normative values that organize modern society is not necessarily to despair of the modern world itself, which has brought with it extraordinary and positive advances. But it is an attempt, like *Sputnik*, to glimpse new perspectives of the modern networked world and then to rejoin the search for ways, like Soviet cybernetics, to harness private power into the service of improving the human condition. A few general comments on the modern world and its networks follow.

Contingency, Failure, Politics

Not only could our networked world have been otherwise—it can still be otherwise today. One of the values of negative histories such as this one is the reminder that most technological projects "fail" or never come to a decisive end (perhaps both failure and repair occur in the long run).[7] The history of technology shows that most technological projects are not consequential at all—at least in the conventional sense. Technological designs are continuously not realized in operable material form and reproducible prototypes, and the social processes that sustain scientific discovery rarely arrive at a clear consensus. The historical record layers documentation of the fossils and footnotes of "dead" media and their iterant afterlives.[8]

This great apparent failure rate in innovations should help shape our considerations of the causes and consequences of modern technologies, such as computer networks. Contingent histories also help focus public debate better than do popular histories of technology that parade about hackers, geniuses, and geeks marching to the Whiggish beats of technological progress. In negative histories, failures, even epic breakdowns, are normal. Astonishing genius, imaginative foresight, and peerless technical wizardry are not enough to change the world. This is one of the lessons of the OGAS experience. Its story places the conventional concepts of technological successes and failures on the wobbly foundations of the accidents of history. The historical record is a cemetery overgrown in short-lived technological futures: stepping off its beaten paths leads us to slow down and take stock before we rush to crown the next generation of technologists as agents of change.

Perhaps the most hopeful reminder to would-be agents of social change is also the hardest: the OGAS team understood that technological reform is also political reform. A well-connected, talented team spent a generation fighting for the political life of a significant project—and those efforts were not enough. Pity the scientists (and popular observers of science) who believe that because we can isolate technical values in our minds, memos, and mathematics, the alchemies of technological development will triumph. Technologies are both artifacts and agents of change—a point that has been made since Max Weber's elective affinities (between Protestantism and capitalism) and Ludwig Fleck's social construction of science.[9] In the multivariable calculus of social reform, the only thing more certain than the injunction that one must try to change the world (and media technologies are one among many ways to do so) is to admit there is no guarantee that any given effort ever will.

A Nod toward Comparative Networks

How does the Soviet case compare to others? The OGAS tale intimates that among the many variables in midcentury network projects—in this case, Soviet socialism, cybernetic science, and decentralized networks—the most important is the institutional environment for technological development. Local institutional behavior is the concrete or quicksand into which the history of networks is poured. Unlike the civilian-oriented Soviet OGAS Project, the Chilean Cybersyn Project, and the (commercial) French Minitel network, the military-initiated U.S. SAGE and ARPANET projects had major effects on civilian industry and society. If there is a virtue to the postwar American military-industrial-academic complex, perhaps it is that the *complex* allowed for cross-sector knowledge exchange and innovation transfer. The failure of the Soviet knowledge base was arguably that the Soviet military consumed resources and hoarded innovations from the civilian economy.

Secondary to that argument, international communication networks precede international computer networks. Without international cybernetic science discourse, the local dialects of systems science in the USSR, Chile, and the United States could have taken different paths and perhaps found design analogies for national networks other than the human mind (for example, the socialist network as a nervous system in the body of the nation and the liberal network as a neural network in the brain of the nation).[10]

The other huge socialist state anchoring the Eurasian steppe makes a good comparison point. The People's Republic of China is, like those states in the former Soviet territories, a socialist state that is now devoted to developing mixed capitalist markets without democracy. Both China and Russia today operate according to informal networks of influence (*guanxi* and *blat*) and are commercializing international computing innovations. The sleek Baidu search, Youku video, and Sina Weibo microblogging platforms imitate and improve the functionalities of Google search, YouTube video, and Twitter. Both states also implement state controls to control national computer network traffic. The most impressive of these is the "great firewall of China," which permits elites and technical experts an escape hatch from the Chinese walled-garden version of the global Internet.

International communication networks also helped to jumpstart and also consign to limbo local computer network projects. This account highlights three case studies: first, Anatoly Kitov's discovery of Norbert Wiener's *Cybernetics* in a secret military library set into motion an internal transition in Soviet scientific discourse; second, Donald Davies and the British Telecom industry prompted the U.S. government to revisit Paul Baran's RAND research on distributed packet-switching networks; and third, news of the ARPANET going online in 1969 prompted the Politburo to revisit the decade-old OGAS proposal in 1970. In each case, international communication networks (even when they were closed or secret) initially prompted internal institutions to revisit concurrent innovations closer to home. As it is in war, so it is in technology: rivals mimic each other mimicking each other. Even so, cold war research networks were evidently too fixated on the international exchange of knowledge among distant friend and foe. Baran openly published his research in the early 1960s, for example, which appears to have delayed his supervisors from attending to his work for several years. Soviet scientists would have discovered Wiener's *Cybernetics* years earlier, and likely to far less sweeping effect, had the book not been banned (and had Stalin not repressed enemy sciences so vigorously). The OGAS proposal probably would have received a fairer public hearing had it not been a secret state project (and had there been a robust Soviet public to share it with). Stretched between contending households that were fueled by the same knowledge anxiety, cold war communication research networks left few researchers with honor in their own lands. In cold war science research, it appears that the more distant and closed the discovery, the easier our narcissism; the closer and more open the discovery in states of emergency, the easier our negligence.

Making Modern Network Culture Strange

The story of the OGAS Project reveals a network culture whose design values—the cybernetic nervous system of the nation, socialist technological utopianism, and decentralized computer networks—now appear to be peculiar to its own time and place. This sustained glance at the strangeness of socialist network projects helps make familiar the foreignness of the modern network culture in historical relief. Consider a hardy perennial of new media thought, the politics of technological utopia, for the OGAS Project was nothing if not a projection of an intrepid socialist future. Socialist politics are no strangers to expansive, sometimes wild flights of imagination about the bounteous blessings of technology. Although technological utopianism belongs to social projects of all types, the socialist tradition boasts a special breed of thinking, including the French socialist utopian thinker Charles Fourier (whose early interests in architecture and engineering were thwarted and who later worked briefly in Paris as head of the Office of Statistics), Karl Marx (who theorized about a socialist revolution near the end of the Industrial Revolution in London), Nasser in Egypt, Tito in Yugoslavia, Nehru in India, the Fabian Society and Labor Party in the United Kingdom, Allende's Cybersyn Project in Chile, and most recently the (independent) Pirate Party of Sweden.[11] In each of these cases, the socialist impulse seeks to flatten out social relations, structurally reorganize society, automate and ease labor, roll out statistical (state) accountability, and gather knowledge that lightens, lifts, and liberates people (even though the effects of such technological utopianism often leans toward shades of dystopia).[12] By imagining the OGAS as a means to a brighter networked Communist future, its architects brought upon the project the full brunt of the *oikos*-led inequalities that drove the administration of Soviet socialism. Perhaps the cardinal mistake of the socialist imagination of technology is not to dream the celebrated dream of social justice but to bulldoze the rutted world of human relations with the private interest logics of the *oikos* (military, corporations, states, and individuals that seek only their own survival).

The Soviet OGAS figured out the "why?" (socialist utopia) but not the "how?" for their large computer network projects, and researchers at the U.S. ARPANET knew the "how?" (packet-switching networks) but not the "why?" of modern networking. The Soviets' missing "how?" lasted for the duration of the project, and the absence of the Western "why?" remains both its historical attraction and the contemporary challenge to computer network culture.

The Western network "how?" has sped many unfinished attempts at answering the network "why?" The technical openness of packet-switching networks to diverse actors has afforded the Internet astonishing and well-documented successes of technical energy, commercial innovation, and cultural creativity. At the same time, the open-ended "why?" that has permitted such generativity has also tolerated the entrance of private forces that are interested in seizing possession of the operating systems and communication infrastructures that mediate the globe. What Arendt observed in the age of *Sputnik* still holds true in the age of smartphones: our technological capacity exceeds our political will to negotiate the terms of that capacity. Our networks are no longer flat (if they ever were) but rather are a consequence of network openness. Our lot, like that of the Soviets, is to live in complex heterarchical power arrangements. Open network cultures are slouching toward tethered devices, nonportable applications, walled gardens (closed platforms), mobile contracts, and much else online and off. At the individual level, these developments further feed and speed the parallel encroachments of private communication forces worldwide, especially the recently documented unprecedented surveillance of national and international communication networks by governments and corporations in the United States and the United Kingdom. Surveillance is the massification of private attention and the antithesis of public attention (the first is a form of global private labor, and the second, personal action).

Two generations ago, a few Soviet actors thought that the OGAS was a good idea. Many more thought that it was a bad idea, and the many won out. A generation ago, many Western observers thought that the Internet was a good thing. The many this time were wrong. The Internet is not a good thing, and it is not a bad thing. It is not a thing at all. The Internet is many things, and many of those things are far less pleasant than cat videos (cat videos feature creatures that, like many human spectators online, enjoy the asocial separation that the screen affords them from their viewers).[13]

This time, however, a few complex private forces are winning out, despite the delusions of digital utopianism or quietism. Whatever else the Internet is (interoperable, generative, nonproprietary, a platform for other platforms), it is not public. As the history of the OGAS indicates, when the public will to confront the high costs of modern network cultures is absent or abused, private forces gladly rush in.

Consider the consequences of this Arendtian argument for modern debate about publicity and privacy. It suggests one way of rereading the term *privacy* in light of rise of the private logics of the *oikos*. Just as the English term *publicity* now belongs to the corporate practice of public relations, so too does the term *privacy* (well before its legal coining as the right to be left alone in 1890) belong to the private concerns of the state and the market, not the person.[14] In this sense modern privacy is not about the proper spacing of the individual self and the other. It is about the sum of private institutional interests that adjudicate the proper spacing of their institutional homes (*oikos*) and the public. War rooms, closed sessions of the Senate, and boardrooms are where modern-day "privacy" resides, in the sense that these are the institutions most interested in "the state of being privy to" the lives of the public. Perhaps due to a mistaken understanding of privacy that emphasizes the individual, not the institution, scholars find the term in "disarray" almost unintelligible outside a particular institutional context, and other languages have trouble translating the English-language lexeme. Perhaps we have misunderstood the term *privacy* all along.[15] It is not what Soviet citizens, under surveillance, never enjoyed. It is the rise of the compulsive power of private forces themselves, which the USSR (among other modern states) was permeated with. Private parties (including *the* Party) and private secretaries (no matter how general or particular) directed organizational forces (however informal, decentralized, and unpredictable) that were bent on securing their own survival at the cost of others. The term *privacy* has not been refeudalized so much as it stands for the colonial expansion of the fiefdoms of institutional power.

Perhaps privacy scholarship should not seek to recover lost individual privacy (the right to control the disclosure of personal information or alternately the right to be left alone) but should critique the malignant growth of institutional privacy (the right to own and its expansion to immortal entities) whether the all-seeing eyes and ears of Google, the National Security Agency, an OGAS-led command economy, or other institutions engaged in massive amounts of information processing (in each example, the economic liberal distinction between private corporation and public state obscures more than it reveals). Glushkov's computer networks would have made the *oikos* of Soviet state-corporation even more privy to the work lives of the Soviet people. In a sense, this is precisely our lot now: the networks that organize *oikos* powers are not hierarchical or decentralized (like the institutions that check them). They are ambiguous, multiple, and heterarchical. They vie for our attention, time, and action. In a time when corporations roam the earth as legal persons, the shadows of Soviet networks are cast on the walls of the present. We might add to Adam Smith's famous warning that businessmen seldom meet without plotting against their consumers: generals, politicians, and the clerisy fare not much better when rolling out the privatizing logic of domination and need. With few exceptions, large networked organizations are inclined to restrain each other only when they interfere with one another in the common race to privatize—or to use—the user. Since before *Sputnik*, our skies, screens, and social lives have been filling with the drones of private network power.

Then and now, the polity and policy landscapes are not identical, and we should not imagine them to be so. The private interests that kept computer networks from being built in the USSR have since hijacked democratic potentials in global networks. The basic institutions that stitch together the social and political fabric of democratic society—the rule of law, functioning courts, equitable tax compliance, Madisonian checks and balances, human and civil rights, an independent press, and private institutions—underlie the often ambiguous and always limited moral foundations of all modern information societies and economies, even informal economies.[16] The patronage socialism of the Soviet Union (like the crony capitalism of modern-day Russia) was missing many of these elements (it had no rule of law, no predictability of procedure, no regulated financial environment, no bankruptcy law, no antirust law, no courts for managing property disputes, and no virtuous regulation of inseparable market and state), but this routine criticism risks ignoring the bigger picture.

Perhaps the choice in the era of cybernetworks has never been between the state and the market as the dominating metaphor for modern networks. We need not accept as final either Glushkov and Cooley's analogy of the state as a nervous system (and the nation as its economic body) or McCulloch and Baran's analogy of the nation-state as a brain (and the network as its neural net). Perhaps the way forward begins with criticizing both cybernetic network analogies for privileging the image of the private mind as supreme. The dominant metaphors for midcentury networked economies—market and state—move us no further than the cybernetic, and ultimately human, hubris that the human mind organizes the world.

Although the landscape between the OGAS Project and the Internet today varies widely, our hopes and despairs pivot on the same things that concerned the creators of the OGAS Project. There is a potential moral authority in institutions and communities to check or caution the large and unscrupulous actors that are intent on networking the world with the creeping private logics of domination. We face new challenges, even as we continue to target the cruelty, corruption, and compulsion of the world that bears us—in every desire to limit the private mind of the *oikos*, there is already a drop of our common human condition.

The OGAS story, therefore, is not only a tale that took place long ago and far away. It can be seen as an allegory of our own lot today. The private forces that were hard at work in the OGAS story are also hard at work in the modern media environment. Informal networks abound, for better and worse. We should not gaze at the OGAS Project from a comfortable distance but realize how close its story hits to home. A world of difference separates all allegories, but looking in the rearview mirror of history, the distance between networked private powers is often closer than it appears.

Coda: A Contingent Legacy of Modest Networks

Beneath the modern imagination of smooth steel-brushed machines interlinked by wires, signals, and smart protocols pulse the vibrant social networks of relations whose virtues and vices have long been part of the human condition. To understand modern networks is at root an exercise in social self-discovery. Our network world shares with the fate of the OGAS Project the vices of self-interest, apathy, back-stabbing, vain imaginations, stupid conceit, *poshlost'* (roughly the "self-satisfied vulgarity" of the petty businessman and administrator, such as Chichikov in Gogol's *Dead Souls*), and all the rest. At the same time, it also shines brightly with generosity, engagement, visionary insight, genius, *byitie* (another untranslatable Russian term meaning roughly "being," "apperception," or a higher state of conscious reality that is resonant with Heideggerian being and scriptural genesis), and much more. The networks binding the human condition can be neither separated nor reconciled. Modern observers can no sooner state the optimal conditions under which humankind has or will best enter the age of global computer networks than we can solve the puzzle of the human condition itself, although the attempts to solve the puzzle are worthwhile. Given that there is no magic solution to these questions, we might do best to seek a modest and cautious perspective on the causes and consequences of the Soviet network experience.

Let us return for a moment to an earlier sense of the word *technology*. In English usage until the early twentieth century, technology was not the hard stuff of tractors and circuits boards but was the study of industrial arts, crafts, and techniques that organize, reveal, and frame the modern world.[17] The suffix -ology in the term technology also appears in the term biology, the study of life. Perhaps by understanding techne- as the artifacts of acculturated human culture (behavior, gesture, oral, literate, print, industrial, mass, and information media and much else), the term technology gives momentum to the study of the crafts of social life.

The Soviet network history teaches several lessons. First, the ambitious, far-seeing faith in the social consequences of technology is no guarantee of technological change in the modern information age. Presentists who look back at the science fiction, fact, and factions of Soviet cybernetics may divine in these pages prophetic prefigurations of modern-day cloud computing, e-commerce, big data processing, and much else. Opportunists may be tempted to enthuse about recuperating the unrealized possibilities of macroprocessing, natural language programming, a self-governing economy, and perhaps even digital immortality, although they will do so in their own tones and cadences. Second, Marx got the point of technology wrong. He wrote that the relations of production—the social relations that all people must enter into in modern life—are fundamental to all else. Another lesson of the OGAS Project is that far more substantial than the hard stuff of technology (cotton mills, industrial factories, hydroelectric dams, nuclear power plants, and the factory and federated computer networks examined here) are the subtle, mundane techniques that continuously work themselves out in the complex relations that constitute being social. Finally, the critic Raymond Williams was right to attend to what might be called the means of sociocultural production, not just the means of industrial production. We can push the point further: the technological means of world production are not just the mass media of newspaper, radio, television, and computer but every commonplace device, understated technique, and learned skill—from a baby's first vocalization to the experienced insider's knowledge of a bureaucracy's peculiarities. These technologies and techniques, creatively read, produce and manage a more genuine base for understanding the arrangement of relations in modern society.[18]

That subtle and modest techniques hold sway over sophisticated information technologies is a clear moral to this story. Letters to leadership found their own random paths in the packet-switching labyrinth of Soviet state. Everything—sudden success, interception and dismissals, evasive telephone calls—came in reply. In fact, the first civilian-military national network proposal anywhere was scuttled because a supervisor did not intercept one letter but did intercept the next (such was the post in the Soviet military). The early institutional alliances between the Central Economic-Mathematical Institute and the Institute of Cybernetics drifted apart over differences about the scale (micro and macro, respectively) at which the mathematical techniques for modeling economic relations should be carried out. The OGAS Project—the ambitions of technocratic economic reform by network—was nearly approved and funded except that two chairs at a committee meeting went unoccupied. National technical networks connecting factories were approved but never realized at the same time that local computer centers in those factories were built but never interconnected—all because of coordination problems (our coordination problems are as great today as their solutions are subtle). Sophisticated chess algorithms outmaneuvered long-term national planning methods and even the occasional chess master, but never to the same effect as a simple notational system kept on index cards (and now online databases). Ministerial ecosystems of paperwork collided and proliferated, and the committee meeting—that omnipresent black box of bureaucracies (even written minutes leave opaque the logics of small-group decisions)—remains among the most undertheorized and delicate techniques governing modern private power networks. Trains and telephone calls were taken and missed; doors opened and locked; hearts and minds pushed to their limits—and sometimes beyond.

The history of Soviet networks showcases something more enduring, powerful, and subtle than a plumbing and sounding out of the stately heights of electronic socialism (although it also does that). It reveals the modest media on which our social relations turn—labyrinthine committee reports and paper trails, bureaucratic and budgetary categories that scrimmage careers, the semantic vagaries of public press releases and precise accounting, empty chairs and scattered letters, accidental meetings in hallways and dachas, and all the other errata of the constant communication and infrequent communion that arrange our lives. When I set out to research the Soviet networks, I hoped for historical insights into the media of tomorrow, but what I found instead were dusty, derelict, and sometimes dispensable residual artifacts of a technological vision for a labyrinthine state now largely forgotten. Not only was I wrong to look for a peek into the future in the archives of the past, I was wrong to think I had not found them. Because the techniques of paper knowledge and print culture continue to accumulate in the scattered anecdotes and artifacts that make up our societies and the stories we tell about them, they too will likely endure as the media of tomorrow. These are the media technologies, writ large, that govern the computer networks and other props of the current information age; theirs are the modern media networks that matter most.

The OGAS Project, like most information age projects, has more of bureaucracy than bits to it. The history and perhaps the future of the current information age will have less to do with the next generation of futurist technologies than it will with the networks of actors and institutions governing the conditions of social relations and the use of knowledge. It would be a mistake to conclude that this far-seeing generation of Soviet scientists and technologists did not realize a network that was capable of changing the world. Their dreams and ambitions were realized not in the networks of steel and silicon chips but in the networks that long have and will continue to govern our lives. The All-State Automated System Project lives on as a story refracted in the records of print culture. In the end, the story told here tells its own moral and method. It asks us to distinguish and extract it from the swirling and glorious strangeness of all scientific ambition that buoys the modern world, exert good will to tolerate it in its oddities, critique it not for what it has not accomplished but for its courting of the irresistible enchantment of modern-day network visions, and finally perhaps even to grow used to it, to wait for it, and to have one day admitted its passage and place into the greater living network of ideas and institutions that make up the modern world.

Such is the uneasy history of Soviet networks. Networks are not the application of a theory of networks, nor are they the children of hard gadgetry and pragmatic engineering. They are the technical arrangements of social relations that have and will continue to change the world. Much remains appropriately and implicitly contingent and unpredictable in the historic making and unmaking of global networks. May the story of the Soviet networks and their troubled paths into an alternative information age stand as sentinel cautions for our networked times. It is not in the nature of daring ideas and the routines of history to come to an end, although such is the lot of books.

Acknowledgments

No book can exist without supporting scholarly and institutional networks, and this is especially true of this book, given its profound debt to mentors, colleagues, friends, and institutions that I engaged with along the way. Whatever commonsense and clear writing is contained in this book—which is an expansion and complete reworking of about half of my Ph.D. dissertation (Columbia University, 2010)—is due to the patient influence of Michael Schudson, a model mentor, adviser, and scholar. My research interests owe Todd Gitlin, Richard John, the late Catharine Nepomnyashchy, and Siva Vaidhyanathan far more than their mentorship. In addition to the four mentors at four schools to whom this book is dedicated (Gary Browning at Brigham Young University, Fred Turner at Stanford University, Michael Schudson at Columbia University, and Joli Jensen at the University of Tulsa), I have to acknowledge my debt to mentors and teachers such as Craig Calhoun, the late James W. Carey, Monika Greenleaf, and Andie Tucher, among many others. At the University of Tulsa, my colleagues Mark Brewin, John Coward, and Joli Jensen have fashioned a work environment collegial enough to make any young professor enviable.

A portion of the first chapter was previously published as the article "Normalizing Soviet Cybernetics" and appears here with permission from the journal *Information & Culture*.

I thank the good folk at my home Ph.D. degree in communications program at the Columbia University Graduate School of Journalism as well as the Archives of the Russian Academy of Sciences, the Institute of Cybernetics of the Ukrainian Academy of Sciences, the Central Economic-Mathematical Institute of the Russian Academy of Sciences, the Institute of Cybernetics of the Ukrainian Academy of Sciences, Elena Vartanova and the School of Journalism at Moscow State University for hosting research visits, the Institute Archives and Special Collections at the Massachusetts

Institute of Technology, and the Freedom of Information Act governing the Central Intelligence Agency and the Federal Bureau of Investigation. I also thank the patient staffs at Butler Library at Columbia University, Widener Library at Harvard University, the Harold B. Lee Library at Brigham Young University, and that modern-day library of Alexandria, interlibrary loan and online scholarly databases. Support and fellowship have come from the Oklahoma Center for the Humanities and faculty summer development grants from the University of Tulsa, the Nevzlin Center for the Study of Russian Jewry and the Lady David Postdoctoral Fellowship at Hebrew University, the Kenneth E. and Becky H. Johnson Foundation, and Junior and other fellowships from the Harriman Institute at Columbia University. Audiences have contributed much at Yale Law School, the Columbia University School of Journalism and the Harriman Institute, the University of Tampere, the Mohyla School of Journalism, the Princeton University Center for Information Technology Policy, and many others. The extraordinary communities in the orbit of Jack Balkin's Information Society Project at Yale Law School and the Berkman Center for Internet and Society at Harvard University have delighted and engaged me for nearly a decade now.

The research animating this book could not exist without the base of work laid by Slava Gerovitch: I am grateful to him for his mentorship. Without the friendship and historical scholarship of Aleksei Viktorovich Kuteinikov and especially the grace and resourcefulness of Vera Viktorevna Glushkova, this introduction to Viktor Glushkov's story to the English-speaking world would likely not exist in book form. I am also grateful to Vladimir Anatolevich Kitov for helpful scholarly resources about his father, Anatoly Kitov, as well. I thank them all three. Previous drafts have benefited from the valuable comments of Geof Bowker, Peter Sachs Collopy, Paul Edwards, Bernard Geoghegan, Lydia Liu, Eden Medina, and Mara Mills on cybernetics and information theory, while Alex Bochannek, Elena Doshlygina, Michael Gordin, Loren Graham, Martin Kragh, Adam Leeds, Ksenia Tatarchenko, and others have taught me much about the Soviet situation. At the risk of leaving many others unnamed, I would also like to thank Colin Agur, Karina Alexanyan, Chris W. Anderson, Mark Andrejevic, Rosemary Avance, Burcu Baykurt, Valerie Belair-Gagnon, Jonah Bossewitch, Gabriella Coleman, Laura DeNardis, Jeffrey Drouin, Maxwell Foxman, Alexander Galloway, Gina Giotta, Abe Gong, Eugene Gorny, Orit Halpern, Lewis Hyde, Andryi Ishchenko, Carolyn Kane, John Kelly, Beth Knobel, Liel Liebovitz, Deborah Lubken, Kembrew McLeod, David Park, Ri Pierce-Grove, Amit Pinchevski, Jefferson Pooley, Erica Robles, Natalia Roudakova, Chris Russil, Jonathan Saunders, Limor Schifman, Trebor Scholz, Steven Schrag, Zohar

Sella, Lea Shaver, Bernhard Siegert, Peter Simonson, Thomas Streeter, Ted Striphas, Patrik Svensson, McKenzie Wark, David Weinberger, and Jonathan Zittrain for helpful conversations and comments on drafts over the years. To me goes the real award of association with them. Menahem Blondheim, Paul Frosh, Elihu and Ruth Katz, Amit Pinchevski, and Limor Shifman, among others, gave me far more than a year of intellectual stimulation and companionship at Hebrew University; Lucas Graves, Rasmus Nielsen, and Julia Sonnevend make ideal conversation partners in our ongoing search for social theoretical understanding. Only two people, to my knowledge, have read the whole manuscript—John Durham Peters and Audra Wolfe. To my delight, both have provided invaluable professional criticism as both scholars and editors; the book is better for the portion of their comments I have incorporated and diminished by those left undone. My editor, Sandra Braman, has buoyed my writing process with her abiding support of the project from start to finish. These are some of the people—alongside the good folk at MIT Press, especially Margy Avery, Deborah Cantor-Adams, and Rosemary Winfield as well as the generous and detailed criticism of two anonymous reviewers—that have improved the book in your hands.

Scholarship began for me first as a family affair. Anyone who knows the depth and breadth of my father's intellectual generosity—a way of life and a man I have long looked up to—should also know that I am at home in my father's field because I am my mother's son. My enduring love and gratitude go to Marsha Paulsen and John Durham Peters for modeling what matters. My own family has made considerable sacrifice and contributions as well. My favorite ruffians and readers—Aaron, Elliot, Libbie, and Maya— have made precious the hours I have spent working on what Libbie has too generously titled "a very long book by Tato Peters" and more precious still the balance of time I have with them. Above all, Kourtney Lambert—my love and favorite reader—makes it all worthwhile.

Appendixes

A   Basic Structure of the Soviet Government

This brief appendix provides a simple outline of the complex and changing structure of the government of the Union of Soviet Socialist Republics (USSR). The country was divided into one federated socialist republic (Russian) and fourteen soviet socialist republics (Armenian, Azerbaijan, Byelorussian, Estonian, Georgian, Kazakh, Kirghiz, Latvian, Lithuanian, Moldavian, Tajik, Turkmen, Ukrainian, and Uzbek). Each republic contained stacked (and sometimes confused) subdivisions ranging from smallest to largest in this order: *raion* (districts, areas, subdistricts), *krai* (territory), *okrug* (district), and *oblast'* (region).

The basic structure of the Soviet state had three parts or political bodies—the Communist Party, the bureaucracy, and the legislature (this ignores the mostly toothless judiciary of the Supreme Court). The Communist Party of the Soviet Union, the only political party permitted by the constitution, coordinated all the affairs of the economy and society. The pyramid party structure rested on a selection of Soviet citizens (no more than 9 percent of the Soviet people were ever members of the Communist Party), and membership was overwhelmingly made up of professional and often technocratic males (the Party shares this with the current digerati demographic). The party structure stretched upward from the members to local party organizations, to local, district, and regional congresses, to the National Party Congress, to the Central Committee, and finally to the Politburo, which was the governing Party committee of the land. At the head of the Politburo sat—in a fitting encapsulation of the Party's bureaucratic spirit—the general secretary, a position that Stalin granted almost supreme powers after Lenin's death. The general secretary worked in theory alongside the premier (the bureaucracy) and the president of state (the legislature) and oversaw the Secretariat, a second ruling Party committee on a level with the Politburo.

The central structure of the bureaucracy scans simply but proved laby-rinthine in practice. At the bottom again were the people, and at the top was the premier, who oversaw the Council of Ministers. Between the citizens and the Council of Ministers fell the internal structures of between twelve and thirty-seven ministries (such as the Ministry of Agriculture and Food and the Ministry of Transport Construction) and the military (the Red Army). During economic reforms, ministries were regularly reorganized, consolidated, and strengthened, and many of them worked across local, district, and national committee subdivisions. This analysis underscores the Soviet bureaucratic divide between civilian ministries and the military (which was a training ground for Party leadership and a sink for the national budget).

Lastly, the legislature was constitutionally appointed in 1918 to oversee economic, social, and security affairs, although in the latter half of the twentieth century its power was largely secondary to the Party and the bureaucracy. Citizen-elected local, district, and regional soviets (or councils) informed the Supreme Soviet, the Presidium, and the president or head of state, whose powers paled in comparison to the premier (head of the bureaucracy) and the general secretary (head of the Party). The Presidium was initially a decision-making body that was a peer with the Council of Ministers (bureaucracy) and the Politburo and Secretariat (Party), although its influence waned with the consolidation and decentralization of power in the Party and bureaucracies under and after Stalin.

These three branches of government were staffed by the *nomenklatura* or elite responsible for higher positions of authority. Formally, the *nomenklatura* occupied a small, elite subset of the already elite Party membership, although in practice it also could include the *intelligentsia* or needed experts who did not have to be Party members (most of the scientists and administrators featured here were members of the Party and often the *intelligentsia*). In the management of the command economy, Party and state hierarchies were separate and overlapping. So although members of the *nomenklatura* could manage a state-owned factory, they also had to have party approval if they were not party members. In such cases, factory directors might report to the local Party secretary as an ordinary Party member, and the Party secretary would report to the director as an employee. In all, this book offers a reminder that in the management of large organizations, especially the Soviet state and economy, the questions of structure and governance are rarely so straightforward as they may appear on paper.

B Annotated List of Slavic Names

For the ease of the English reader, the text refers to people who recur in this history by first and last names; other persons, no matter how significant, whose names do not appear in the text frequently are named in the Soviet academic tradition of two initials (the first name and patronymic) followed by last name. Only recurring figures are listed below.

Aksel Berg (1893–1973): Engineer admiral, deputy chair of the Council on Cybernetics.

Mikhail Botvinnik (1911–1995): Soviet international grandmaster, founding member of the Soviet school of chess, professional electrical engineer, computer scientist, and champion of early computer chess Pioneer program, and author of several proposals to computerize strategic planning.

Leonid Brezhnev (1906–1982): General secretary of the Union of Soviet Socialist Republics (1964–1982).

Nikolai Fedorenko (1917–2006): Chemist and economist, director of the Central Economic-Mathematical Institute (1963–1985), coauthor of the EGSVTs (Unified State Network of Computing Centers) network project (1963), academician.

Vasily Garbuzov (1911–1985): Minister of finances (1965–1980), principal opponent to the OGAS (All-State Automated System) Project, rival of Vladimir Starovsky and the Central Statistical Administration.

Viktor Glushkov (1923–1982): Prominent Soviet cyberneticist, director of the Institute for Cybernetics in Kiev, Ukraine (1967–1982), author of OGAS (All-State Automated System) (1963–1982), coauthor of the EGSVTs (Unified State Network of Computing Centers) (1963) network projects, academician.

Mikhail Gorbachev (1931–): General secretary, Union of Soviet Socialist Republics (1985–1991).

Leonid Kantorovich (1912–1986): Soviet economic mathematician, pioneer in linear modeling, Nobel Prize in economics (1975).

Mstislav Keldysh (1911–1978): Mathematician, Soviet space theorist, chair Soviet Academy of Sciences (1961–1975) (where he helped rehabilitate cybernetics and genetics).

Aleksandr Kharkevich (1904–1965): Communication engineer, director of the Institute for Information Transmission Problems (1962–1965), author of the ESS (Unified Communication System) network project (1963).

Nikita Khrushchev (1894–1971): First (general) secretary of the Union of Soviet Socialist Republics (1953–1964).

Anatoly Kitov (1920–2005): Mathematician, colonel engineer, first Soviet cyberneticist, coauthor *The Basic Features of Cybernetics* (1955), author of the EASU (Economic Automatic Management System) network proposal (1959).

Ernst Kolman (1892–1979): Failed mathematician, philosopher-critic, accuser of Andrei Kolmogorov (1939), author of "What Is Cybernetics?" (1955), first ideological supporter of Soviet cybernetics (1955–1979).

Andrei Kolmogorov (1903–1987): Prominent mathematician, public cybernetics supporter (1960–1970).

Aleksei Kosygin (1904–1980): Premier of the Union of Soviet Socialist Republics (1964–1980), deputy chair of the Soviet Council of Ministers, appointed Viktor Glushkov and Nikolai Fedorenko to develop the OGAS Project and the EGSVTs (Unified State Network of Computing Centers) network project (1962).

Aleksei Lyuapunov (1911–1973): Mathematician, pioneering cyberneticist, coauthor of "Basic Features of Cybernetics" (1955).

Vasily Nemchinov (1894–1964): Economic mathematician, organizer of the laboratory in Novosibirsk (1958) that became Nikolai Fedorenko's Central Economic-Mathematical Institute in Moscow (1963).

Konstantin Rudnev (1911–1980): Author of a 1963 *Izvestia* article in favor of using computers in national planning, head of the Ministry of Instrument Making, Automated Equipment, and Control Systems (1965–1980).

Sergei Sobolev (1908–1989): Prominent mathematician, coauthor of "The Basic Features of Cybernetics" (1955), public supporter of cybernetics (1955–1970).

Vladimir Starovsky (1905–1975): Director of the Central Statistical Administration in the Council of Ministers (1957–1975), principal opponent of the OGAS (All-State Automated System) Project, rival of Vazily Garbuzov and his Ministry of Finance.

C Network and Other Project Acronyms

ARPANET Advanced Research Projects Agency Network, later Defense Advanced Research Projects Agency Network (1969–1983), United States Department of Defense, first packet-switching network and predecessor to the Internet.

ASU (*avtomatizirovannaya sistema upravleniya*): Automated system of management. The Soviet term for a management information and control system, or, effectively, a local network between an onsite computer and attending industrial processes that it supervises at a factory.

CEMI (*tsentralnyi ekonomicheskii-mathematicheskii institute*): The Central Economic-Mathematical Institute of the Academy of Sciences in Moscow, proposed by Nemchinov, built on his Laboratory of Economic Mathematical Methods, founded in 1963, first directed by Nikolai Fedorenko, and an early collaborator with Viktor Glushkov's Institute of Cybernetics on the OGAS (All-State Automated System) Project and the EGSVTs (Unified State Network of Computing Centers) network projects.

CSA (*tsentral'noe statisticheskoe upravleniye*): The Central Statistical Administration (or Directorate) was, as part of the Council of Ministers (the highest executive council in the Soviet Union) between 1948 and 1987, the main organization in the Soviet state charged with statistical oversight.

EASU (*ekonomicheskaya avtomatizirovannaya sistema upravleniya*): Economic Automated Management System proposed by Anatoly Kitov (1959).

ESS (*edinaya sistema svyazi*): Unified Communication System, a comprehensive data communication network planned by Aleksandr Kharkevich (1963).

EGSVTs (*edinogosudarstvennaya set' vyichisletel'nikh tsentrov*): Unified State Network of Computing Centers, technical base of the OGAS (All-State

Automated System) Project, coauthored by Viktor Glushkov and Nikolai Fedorenko (1963). In other literature, associated with a complex series of other subdevelopments. The EGSVTs was a subset of the overall OGAS Project.

OGAS(U) (*obshche-gosudarstvennaya avtomatizirovannay sistema upravleniya*): All-State Automated System (of Management). Inspired by Anatoly Kitov's EASU (Economic Automatic Management System) and composed of a national network connecting and managing ASUs (automated system of management), it was proposed by Viktor Glushkov and others between 1963 and 1985, developed variously by the Institute of Cybernetics, CEMI (Central Economic-Mathematical Institute), and others. EGSVTs (Unified State Network of Computing Centers) was projected to be its the technical base. SOFE (System for the Optimal Functioning of the Economy) was projected to be its modeling system.

SAGE: Semi-Automatic Ground Environment, an air defense control system used by the United States and Canada from the late 1950s through the 1980s. Although ineffectual as a strategic network, it appears to have been an important site for developing online, real-time interactive computing over long distances.

SOFE (*sistema optimal'nogo funktsionirovaniya ekonomiki*): System for the Optimal Functioning of the Economy, developed under Nikolai Fedorenko at CEMI (Central Economic-Mathematical Institute), which pioneered systems models and theories for optimizing economic planning since the 1960s. Initially a companion program for developing the optimization and economic management software behind the OGAS (All-State Automated System) Project.

Notes

Prologue

1. Flo Conway and Jim Siegelman, *Dark Hero of the Information Age: In Search of Norbert Wiener, the Father of Cybernetics* (New York: Basic Books, 2005), 392 n. 318.

2. Marshall McLuhan, *Understanding Media: The Extensions of Man* (New York: McGraw-Hill, 1964).

3. Slava Gerovitch, "InterNyet: Why the Soviet Union Did Not Build a Nationwide Computer Network," *History and Technology* 24 (4) (December 2008): 335–350.

4. Viktor Shklovsky, "Art as Technique" (1917), in *Russian Formalist Criticism: Four Essays*, ed. Lee T. Lemon and Marion J. Reiss (Lincoln: University of Nebraska Press, 1965), 3–24.

5. Peter Brown, *The Body and Society: Men, Women, and Sexual Renunciation in Early Christianity* (New York: Columbia University Press, 1988), xvii.

Introduction

1. On September 19, 1990, fifteen months before the Soviet Union collapsed, the Internet Corporation for Assigned Names and Numbers (ICANN) assigned the .su country code top-level domain, and it remains in use today.

2. For more on Akademgorodok, see Paul R. Josephson, *New Atlantis Revisited: Akademgorodok, the Siberian City of Science* (Princeton: Princeton University Press, 1977).

3. The literature on the Soviet Union's role in the cold war is enormous. Readers unacquainted with that literature may wish to start with a primer on the global cold war context, such as Robert J. McMahon, *The Cold War: A Very Short Introduction* (New York: Oxford University Press, 2003), Steven Lovell, *The Soviet Union: A Very Short Introduction* (New York: Oxford University Press, 2009), and a more substantial

work by Orlando Figes, *Revolutionary Russia, 1891–1991: A History* (New York: Metropolitan Books, 2014). Other classics outside the Soviet period or space include Eric Hobsbawm, *The Age of Extremes: A History of the World, 1914–1991* (New York: Pantheon Books, 1994); Orlando Figes, *Natasha's Dance: A Cultural History of Russia* (New York: Picador, 2003); and James H. Billington, *The Icon and the Axe: An Interpretive History of Russian Culture* (New York: Vintage, 1966). For more on the intellectual context, see the politically opposing pair, Isaiah Berlin, *Russian Thinkers* (New York: Penguin Group, 1978), and Richard Pipes, *Russian Conservatism and Its Critics: A Study in Political Culture* (New Haven: Yale University Press, 2005).

4. Robert E. Kohler and Kathryn M. Olesko, "Introduction: Clio Meets Science: The Challenges of History," *Osiris* 27 (1) (2012): 4–6.

5. The literature on the history of computing in the United States context is also significant. For a basic introduction, see Paul E. Ceruzzi, *Computing: A Concise History* (Cambridge: MIT Press, 2012); Paul E. Ceruzzi, *A History of Modern Computing* (Cambridge: MIT Press, 1998); Martin Campbell-Kelly and William Aspray, *Computer: A History of the Information Machine* (Boulder, CO: Westview, 2004); and William Aspray and Paul E. Ceruzzi, *The Internet and American Business* (Cambridge: MIT Press, 2008). The growing literature on the U.S. history of the Internet includes works such as Janet Abbate, *Inventing the Internet* (Cambridge: MIT Press, 1999); Paul N. Edwards, *The Closed World: Computers and the Politics of Discourse in Cold War America* (Cambridge: MIT Press, 1996); Finn Burton, *Spam: A Shadow History of the Internet* (Cambridge: MIT Press, 2013); and Thomas Streeter, *The Net Effect: Romanticism, Capitalism, and the Internet* (New York: New York University Press, 2011). See also Jonathan Zittrain, *The Future of the Internet, and How to Stop It* (New Haven: Yale University Press, 2008), and Tim Wu, *The Master Switch: The Rise and Fall of Information Empires* (New York: Atlantic Books, 2010).

6. Scholarship has not yet advanced a deep understanding of the relationship between social justice and computing, although initial inroads are being made in the critical study of gender and computing. A few works of note include Donna Haraway, *Simians, Cyborgs and Women: The Reinvention of Nature* (New York: Routledge, 1991); Jennifer S. Light, "When Computers Were Women," *Technology and Culture* 40 (3) (1999): 455–483; Nathan Ensmenger, *The Computer Boys Take Over: Computers, Programmers, and the Politics of Technical Expertise* (Cambridge: MIT Press, 2010); and Mette Bryld and Nina Lykke, *Cosmodolphins: Feminist Cultural Studies of Technology, Animals and the Sacred* (New York: Zed Books, 2000).

7. David E. Hoffmann, *The Dead Hand: The Untold Story of the Cold War Arms Race and Its Dangerous Legacy* (New York: Random House, 2009), 150–154, 364–369, 422–423, 477.

8. Ibid., 153–154.

9. For sample references, see Kevin Kelly, *Out of Control: The New Biology of Machines, Social Systems, and the Economic World*, Fourth Edition (Reading, MA: Addison Wesley, 2004), chap. 4; Eric Raymond, *The Cathedral and the Bazaar: Musings on Linux and Open Source by an Accidental Revolutionary* (New York: O'Reilly, 1999); and Leon Trotsky, *Platform of the Joint Opposition* (1927) (London: New Park Publications, 1973), especially "The Agrarian Question and Social Construction."

10. Manuel Castells, *End of the Millennium: The Information Age—Economy, Society, and Culture* (Malden, MA: Blackwell, 1998), 5–68; Lawrence Lessig, *Code and Other Laws of Cyberspace* (New York: Basic Books, 1999), 3–8.

11. Yochai Benkler, *The Wealth of Networks: How Social Production Transforms Markets and Freedom* (New Haven: Yale University Press, 2006).

12. Melvin Kranzberg, "Technology and History: 'Kranzberg's Laws,'" *Technology and Culture* 27 (3) (1986): 544–560.

13. For Latour's aphorism, see Bruno Latour, "Technology Is Society Made Durable," in *A Sociology of Monsters: Essays on Power, Technology and Domination*, ed. John Law, Sociological Review Monograph No. 38 (London: Routledge, 1991), 103–132. For an excellent bibliographical bridge between science and technology studies (STS) and the study of information technologies, see P. Boczkowski and L. Lievrouw, "Bridging STS and Communication Studies: Scholarship on Media and Information Technologies," in *The Handbook of Science and Technology Studies*, ed. E. Hackett, O. Amsterdamska, M. Lynch, and J. Wajcman, 3rd ed. (Cambridge: MIT Press, 2007), 949–977.

14. Geoffrey C. Bowker and Leigh Starr, *Sorting Things Out: Classification and Its Consequences* (Cambridge: MIT Press, 1999), 33–50.

15. Eric Hobsbawm, *How to Change the World: Reflections on Marx and Marxism* (New Haven: Yale University Press, 2011), 22–41.

16. The article that made this book possible is Slava Gerovitch, "InterNyet: Why the Soviet Union Did Not Build a Nationwide Computer Network," *History and Technology*, 24 (4) (2008): 335–350. See also Slava Gerovitch, "The Cybernetics Scare and the Origins of the Internet," *Baltic Worlds* 2 (1) (2009): 32–38; Slava Gerovitch, *From Newspeak to Cyberspeak: A History of Soviet Cybernetics* (Cambridge: MIT Press, 2002); Slava Gerovitch, "Speaking Cybernetically: The Soviet Remaking of an American Science," Ph.D. diss., Program in Science, Technology and Society, Massachusetts Institute of Technology, 1999; Loren R. Graham, *Science, Philosophy, and Human Behavior in the Soviet Union* (New York: Columbia University, 1987); Loren R. Graham, *Science in Russia and the Soviet Union: A Short History* (New York: Cambridge University Press, 1993); and Loren R. Graham, *Lonely Ideas: Can Russia Compete?* (Cambridge: MIT Press, 2013).

17. Classic and recent histories of the Internet and its American milieu include Abbate, *Inventing the Internet*; Edwards, *The Closed World*; Burton, *Spam*; and Thomas

Streeter, *The Net Effect: Romanticism, Capitalism, and the Internet* (New York: New York University Press, 2011). For more popular introductions, see Ian F. McNeely with Lisa Wolverton, *Reinventing Knowledge: From Alexandria to the Internet* (New York: Norton, 2008), whose scholarly breadth and snap counterweight popular accounts such as Katie Hafner, *Where Wizards Stay Up Late: The Origins of the Internet* (New York: Simon & Schuster, 1996), and Walter Isaacson, *The Innovators: How a Group of Hackers, Geniuses, and Geeks Created the Digital Revolution* (New York: Simon & Schuster, 2014).

18. I owe a version of this line and much else to conversations with Elihu Katz at the Department of Communication at Hebrew University in the spring of 2011.

19. Eden Medina, *Cybernetic Revolutionaries: Technology and Politics in Allende's Chile* (Cambridge: MIT Press, 2011).

20. The literature on cybernetics, viewed in its breadth, is considerable and growing. For a brief introduction, see Bernard Geoghegan and Benjamin Peters, "Cybernetics," in *The John Hopkins Guide to Digital Media*, ed. Marie-Laure Ryan, Lori Emerson, and Benjamin J. Robertson (Baltimore: John Hopkins University Press, 2014), 109–112. For more on cybernetics in the United States, see Peter Galison, "The Ontology of the Enemy: Norbert Wiener and the Cybernetic Vision," *Critical Inquiry* 21 (1) (1994): 228–266; Geoffrey C. Bowker, "How to Be Universal: Some Cybernetic Strategies, 1943–1970," *Social Studies of Science* 23 (1993): 107–127; Geoffrey Bowker, "The Empty Archive: Cybernetics and the 1960s," in *Memory Practices in the Sciences* (Cambridge: MIT Press, 2006); Lily E. Kay, "Cybernetics, Information, Life: The Emergence of Scriptural Representations of Heredity," *Configurations* 5 (1) (1997): 23–91.Books on the cybernetic context before and during the U.S. cold war include Edwards, *The Closed World*; David Mindell, *Between Human and Machine: Feedback, Control, and Computing before Cybernetics* (Baltimore: John Hopkins Press, 2002); Jennifer Light, *From Warfare to Welfare: Defense Intellectuals and Urban Problems in Cold War America* (Baltimore: Johns Hopkins University Press, 2003); and Darren Tofts, Annemarie Jonson, and Alessio Cavallaro, eds., *Prefiguring Cyberculture: An Intellectual History* (Cambridge: MIT Press, 2002).A few biographical works include Steve J. Heims, *The Cybernetics Group* (Cambridge: MIT Press, 1991); Steve J. Heims, *John von Neumann and Norbert Wiener: From Mathematics to the Technologies of Life and Death* (Cambridge: MIT Press, 1982); Pesi R. Masani, *Norbert Wiener, 1894–1964* (Boston: Birkhäuser Verlag, 1990); Flow Conway and Jim Siegelman, *Dark Hero of the Information Age: In Search of Norbert Wiener, the Father of Cybernetics* (New York: Basic Books, 2005); and Hunter Crowther-Heyck, *Herbert A. Simon: The Bounds of Reason in Modern America* (Baltimore: Johns Hopkins University Press, 2005).A few key theorizations and historical treatments include N. Katherine Hayles, *How We Became Posthuman: Virtual Bodies in Cybernetics, Literature, and Informatics* (Chicago: University of Chicago Press, 1999); Jean-Pierre Dupuy, *The Mechanization of the Mind: The Origins of Cognitive Science*, trans. M. B. DeBevoise (Princeton: Princeton University Press, 2000; Cambridge: MIT Press, 2009); John Johnston, *The Allure of Machinic Life: Cybernetics,*

Artificial Life, and the New AI (Cambridge: MIT Press, 2008); Philip Mirowski, *Machine Dreams: Economics Becomes a Cyborg Science* (New York: Cambridge University Press, 2001); Orit Halpern, "Dreams for Our Perceptual Present: Archives, Interfaces, and Networks in Cybernetics," *Configurations* 13 (2007): 283–319; Stuart Umpleby, "A History of the Cybernetics Movement in the United States," *Journal of the Washington Academy of Sciences* 91 (2005): 54–66; Bernard Geoghegan, "The Historiographic Conceptualization of Information: A Critical Survey," *IEEE Annals of the History of Computing* 30 (2008): 66–81.For more on cybernetics in the Soviet Union, see Slava Gerovitch, *From Newspeak to Cyberspeak: A History of Soviet Cybernetics* (Cambridge: MIT Press, 2002); David Holloway, "Innovation in Science: The Case of Cybernetics in the Soviet Union," *Science Studies* 4 (1974): 299–337; and David Mindell, Jerome Segal, and Slava Gerovitch, "From Communications Engineering to Communications Science: Cybernetics and Information Theory in the United States, France, and the Soviet Union," in *Science and Ideology: A Comparative History*, ed. Mark Walker, 66–96 (New York: Routledge, 2003).Work on cybernetics in France includes, among others, Celine Lafontaine, "The Cybernetic Matrix of 'French Theory,'" *Theory, Culture and Society* 24 (2007): 27–46; Lydia Liu, "The Cybernetic Unconscious: Rethinking Lacan, Poe, and French Theory," *Critical Inquiry* 36 (2010): 288–320; Bernard Geoghegan, "From Information Theory to French Theory: Jakobson, Lévi-Strauss, and the Cybernetic Apparatus," *Critical Inquiry* 38 (2011): 96–126. On cybernetics in Britain, see Andrew Pickering, *The Cybernetic Brain: Sketches of Another Future* (Chicago: University of Chicago Press, 2010).On cybernetics in East Germany, see Jérôme Segal, "L'introduction de la cybernétique en R.D.A. rencontres avec l'idéologie marxiste," *Science, Technology and Political Change: Proceedings of the Twentieth International Congress of History of Science* (Liège, July 20–26, 1997) (Brepols: Turnhout, 1999), 1: 67–80.And on cybernetics in China, see Susan Greenhalgh, "Missile Science, Population Science: The Origins of China's One-Child Policy," *China Quarterly* 182 (2005): 253–276. On cybernetics in Chile, see Medina, *Cybernetic Revolutionaries.*

21. I owe the term *knowledge base* to conversations with Richard John in 2010. See, in particular, his related work on the political decisions that have shaped U.S. communication history, *Network Nation: Inventing American Telecommunications* (Cambridge: Harvard University Press, 2010).

22. Stephen Jay Gould, *Life's Grandeur* (London: Vintage, 1997), 7.

23. Under the name "actor-network theory," Bruno Latour has attempted to theorize the concept of *network* as a way of retooling the historian's method of following the linkages across all forms of actors. See Bruno Latour's *Science in Action: How to Follow Scientists and Engineers through Society* (Milton Keynes: Open University Press, 1987). Two decades later, he deemed "the word *network* so ambiguous we should have abandoned it long ago," in Bruno Latour, *Reassembling the Social: An Introduction to Actor-Network-Theory* (Oxford: Oxford University Press, 2005), 129–130.

24. For more on the historical designator *new media*, see Benjamin Peters, "And Lead Us Not into Thinking the New Is New: A Bibliographic Case for New Media History," *New Media and Society* 11 (1–2) (2009): 13–30.

25. Aleksandr Ya. Khinchin, "Teoria prosteishego potoka" (Mathematical Methods of the Theory of Mass Service; more literally, Simple Stream Theory), *Trudy Matematicheskogo Instituta Steklov.* 49 (1955): 3–122.

26. János Kornai, *The Socialist System: The Political Economy of Communism* (Princeton: Princeton University Press, 1992); David Graeber, *Debt: The First Five Thousand Years* (New York: Melville House, 2011), 94.

27. The field of institutional economics offers pragmatic approaches to observed irrationalities in individual and group actions. A few standard references in the literature include Thorsten Veblen's heterodox position in "Why Is Economics Not an Evolutionary Science?," *Quarterly Journal of Economics* 12 (1898): 373–393; Thomas C. Schelling, *Micromotives and Macrobehavior* (New York: Norton, 1978); Douglass C. North, *Institutions, Institutional Change and Economic Performance* (New York: Cambridge University Press, 1998); Ronald Coase, "The New Institutional Economics," *American Economic Review* 88 (2) (1998): 72–74; and William Kapp, *The Foundations of Institutional Economics* (New York: Routledge, 2011). For comparison to the quirkiness of individual decisions, see popular introductions to cognitive psychology and behavioral psychology and economics, such as Daniel Kahnemann, *Thinking Fast and Slow* (New York: Farrar, Straus, and Giroux, 2011), and Dan Ariely, *Predictably Irrational: The Hidden Forces That Shape Our Decisions* (New York: HarperCollins, 2008). Compare these to recent works on the informal and violent character of post-Soviet economics, including Alena V. Ledeneva, *Russia's Economy of Favors: Blat, Networking and Information Exchange* (New York: Cambridge University Press, 1998), and Vadim Volkov, *Violent Entrepreneurs: The Use of Force in the Making of Russian Capitalism* (Ithaca: Cornell University Press, 2002).

28. The English-language literature on tech entrepreneurs is long and popular, including Walter Isaacson, *The Innovators* (New York: Simon & Schuster, 2014), and Peter Thiel, *Zero to One: Notes on Startups, or How to Build the Future* (New York: Crown Business, 2014), but very little of it to my knowledge looks beyond the West (in particular, the west coast of the United States and the eastern Asian rim), such as Eden Medina, ed., *Beyond Imported Magic: Essays on Science, Technology, and Society in Latin America* (Cambridge: MIT Press, 2014).

Chapter 1: A Global History of Cybernetics

1. See note 20 on cybernetic literature in the introduction to this book.

2. See Wiener, *Cybernetics*; Bowker, "How to Be Universal: Some Cybernetic Strategies, 1943–70"; Galison, "The Ontology of the Enemy"; and J. R. Pierce, "The Early

Days of Information Theory," *IEEE Transactions on Information Theory* 19 (1) (1973): 3–8; and especially Ronald R. Kline, *The Cybernetics Moment, Or Why We Call Our Age the Information Age* (Baltimore, MD: Johns Hopkins University Press, 2015).

3. Ronald R. Kline, "Where Are the Cyborgs in Cybernetics?," *Social Studies of Science* 39 (3) (2009): 331–362.

4. Wiener, *Cybernetics*. On the curious father-son circularities between Leo's Slavic studies and Norbert's cold war cybernetics, see Benjamin Peters, "Toward a Genealogy of a Cold War Communication Science: The Strange Loops of Leo and Norbert Wiener," *Russian Journal of Communication* 5 (1) (2013): 31–43.

5. This section draws on my previously published work on cybernetics, including Bernard Geoghegan and Benjamin Peters, "Cybernetics" in *The John Hopkins Guide to Digital Media*, ed. Marie-Laure Ryan et. al. (Baltimore: John Hopkins University Press, 2014), 109–112.

6. Wiener's classic works include his technical masterpiece *Cybernetics*, the popular *The Human Use of Human Beings: Cybernetics and Society* (Boston: Houghton Mifflin, 1950), and his deathbed lectures *God and Golem, Inc.: A Comment on Certain Points Where Cybernetics Impinges on Religion* (Cambridge: MIT Press, 1964).

7. Wiener, *Cybernetics*, 1–25, 155–168.

8. Ibid., 16.

9. Dupuy, *Mechanization of the Mind*. See also John von Neumann, *The Computer and the Brain*, 2nd ed. (New Haven: Yale University Press, [1958] 2000).

10. Quoted in Claus Pias, "Analog, Digital, and the Cybernetic Illusion," *Kybernetes* 34 (3–4) (2005): 544.

11. Claus Pias, ed., *Cybernetics-Kybernetik 2: The Macy-Conferences 1946–1953* (Berlin: Diaphanes, 2004).

12. Steve J. Heims, *The Cybernetics Group* (Cambridge: MIT Press, 1991).

13. Ibid., 52–53, 207.

14. William Aspray, *John von Neumann and the Origins of Modern Computing* (Cambridge: MIT Press, 1990).

15. David Lipset, *Gregory Bateson: The Legacy of a Scientist* (New York: Prentice Hall, 1980). See also Fred Turner, *From Counterculture to Cyberculture* (Chicago: University of Chicago Press, 2006), 121–125.

16. Jefferson Pooley, "An Accident of Memory: Edward Shils, Paul Lazarsfeld and the History of American Mass Communication Research," Ph.D. diss., Columbia University, New York, 2006.

17. For more on "trading zones," see Peter Galison, *Image and Logic: A Material Culture of Microphysics* (Chicago: University of Chicago Press, 1997), 44–47, 781–784, 806–807, 816–817.

18. Bill Aspray, "The Scientific Conceptualization of Information," *Annals of the History of Computing* 7 (2) (1985): 117–140.

19. Claude E. Shannon, "A Mathematical Theory of Communication," *Bell Systems Technical Journal* 27 (1948): 379–423, 623–656.

20. Mirowski, *Machine Dreams*.

21. Paul Erickson, Judy L. Klein, Lorraine Dastone, Rebecca Lemov, Thomas Sturm, and Michael D. Gordin, *How Reason Almost Lost Its Mind: The Strange Career of Cold War Rationality* (Chicago: University of Chicago Press, 2013).

22. Claude E. Shannon, "The Bandwagon," *IRE Transactions on Information Theory* 2 (1) (1956): 3. See also Pierce, "The Early Days of Information Theory"; Norbert Wiener, "What Is Information Theory?," *IRE Transactions on Information Theory* 48 (1956): 48; Ronald R. Kline, "What Is Information Theory a Theory Of? Boundary Work among Scientists in the United States and Britain during the Cold War," in *The History and Heritage of Scientific and Technical Information Systems: Proceedings of the 2002 Conference, Chemical Heritage Foundation*, ed. W. Boyd Rayward and Mary Ellen Bowden, 15–28 (Medford, NJ: Information Today, 2004).

23. Arturo Rosenblueth, Norbert Wiener, and Julian Bigelow, "Behavior, Purpose, and Teleology," *Philosophy of Science* 10 (1943): 18–24.

24. Daniel Kahneman and Amos Tversky, "Prospect Theory: An Analysis of Decisions under Risk," *Econometrica* 47 (2) (1979): 263–291. See also Daniel Kahneman and Amos Tversky, eds., *Choices, Values and Frames* (New York: Cambridge University Press and Russell Sage Foundation, 2000).

25. David Stark, *The Sense of Dissonance: Accounts of Worth in Economic Life* (Princeton: Princeton University Press, 2009), 1–34.

26. The intellectual history of thought on hierarchy and its critics would fill many shelves. That history might combine thinking on technical subordination in mathematics (cardinal numbers, graphs, networks, sets, type theory, programming) and other classificatory systems; individual autonomy (Plato, Locke and Kant, Isaiah Berlin and Charles Taylor) and sociobiological evolution; legal, ethical, and religious thought; and pragmatism and feminism. For a helpful update on modern network discourse, see Daniel Kreiss, Megan Finn, and Fred Turner, "The Limits of Peer Production: Some Reminders from Max Weber for the Network Society," *New Media and Society* 13 (2) (2011): 243–259.

27. Warren S. McCulloch, "A Heterarchy of Values Determined by the Topology of Nervous Nets," *Bulletin of Mathematical Biophysics* 7 (1945): 89–93.

28. Ibid., 91.

29. George Dyson, *Turing's Cathedral: The Origins of the Digital Universe* (New York: Pantheon Books, 2012), 196–197, see also 7–10, 56–63.

30. John von Neumann, "Can We Survive Technology?," *Fortune* (June 1955): 106–108, 151–152.

31. For a lively discussion, see Dupuy, *The Mechanization of Mind*.

32. For a few examples of the French scholarly and popular presses on cybernetics between 1946 and 1952, see Jacque Bergier, "Un plan général d'automatisation des industries," *Les Lettres françaises* (April 15, 1948): 7–8; Léon Brillouin, "Les machines américaines," *Annales des Télécommunications* 2 (1947): 331–346; Léon Brillouin, "Les grandes machines mathématiques américaines," *Atomes* 2 (21) (1947): 400–404; Louis de Broglie, *La cybernétique: théorie du signal et de l'information* (Paris: Edition de la Revue d'Optique Théorique et Instrumentale, 1951); Dominique Dubarle, "Une nouvelle science: la cybernétique—vers la machine à gouverner?," *Le Monde*, December 28, 1948, in P. Breton, *A l'image de l'homme* (Paris: Seuil, 1995), 137–138; Dominique Dubarle, "Idées scientifiques actuelles et domination des faits humains," *Esprit* 9 (18) (1950): 296–317. See also Jérôme Segal, *Le zéro et le un: histoire de la notion scientifique d'information* (Paris: Syllepse, 2003).

33. Mindell, Segal, and Gerovitch, "From Communications Engineering to Communications Science."

34. Ibid. See also Geoghegan, "From Information Theory to French Theory"; Céline LaFontaine, "The Cybernetic Matrix of French Theory"; and LaFontaine, *L'empire cybernétique: des machines à penser à la pensée machine* (Paris: Seuil, 2004).

35. Phil Husbands and Owen Holland, "The Ratio Club: A Hub of British Cyberneticists," in *The Mechanical Mind in History*, ed. P. Husbands, O. Holland, and M. Wheeler, 91–148 (Cambridge: MIT Press, 2008).

36. Pickering, *The Cybernetic Brain*.

37. Stafford Beer, *Brain of the Firm* (London: Allen Lane, Penguin Press, 1972).

38. Humberto Maturana and Francisco Valera, *Autopoiesis and Cognition: The Realization of the Living* (Boston: Reidel, 1980); Francisco Valera, *The Tree of Knowledge: The Biological Roots of Human Understanding* (Boston: Shambhala Press, 1987); Francisco Valera with Evan Thompson and Eleanor Rosch, *The Embodied Mind: Cognitive Science and Human Experience* (Cambridge: MIT Press, 1991).

39. For more on Aleksandr Bogdanov, see his *Tektologia: Vsyeobshcheiye Organizatsionnaya Nauka* (*Tectology: Universal Organizational Science*) (Moscow: Akademia Nauk, 1913–1922). See also Nikolai Krementsov, *A Martian Stranded on Earth: Alexander Bogdanov, Blood Transfusions, and Proletarian Science* (Chicago: University of Chicago Press, 2011), and J. Biggart, P. Dudley, and F. King, eds., *Alexander Bogdanov and the*

Origins of Systems Thinking in Russia (Brookfield, VT: Ashgate, 1998), and McKenzie Wark, *Molecular Red: A Theory for the Anthropocene* (New York: Verso, 2015).

40. On Stefan Odobleja, see Mihai Draganescu, *Odobleja: Between Ampère and Wiener* (Bucharest: Academia Republicii Socialiste Romania, 1981); Nicolae Jurcau, "Two Specialists in Cybernetics: Stefan Odobleja and Norbert Wiener, Common and Different Features," *Twentieth World Congress of Philosophy* (1998), accessed October 11, 2011, http://www.bu.edu/wcp/Papers/Comp/CompJurc.htm.

41. Peters, "Toward a Genealogy of a Cold War Communication Science."

42. Michael O'Shea. *The Brain: A Very Short Introduction* (New York: Oxford University Press, 2005), 1.

43. For canonic works on Soviet science written during or soon after the cold war, see Zhores Medvedev, *Soviet Science* (New York: Norton, 1978); Alexander Vucinich, *Empire of Knowledge: The Academy of Sciences of the USSR (1917–1970)* (Berkeley: University of California Press, 1984); Graham, *Science in Russia and the Soviet Union*; David Joravsky, *Soviet Marxism and Natural Science, 1917–1932* (New York: Columbia University Press, 1971). For more current materials, see Nikolai Krementsov, *Stalinist Science* (Princeton: Princeton University Press, 1996); Paul R. Josephson, *Totalitarian Science and Technology* (Atlantic Highlands, NJ: Humanities Press, 1996); Paul R. Josephson, *Red Atom: Russia's Nuclear Power Program from Stalin to Today* (Pittsburg: University of Pittsburg Press, 2005); and Ethan Pollock, *Stalin and the Soviet Science Wars* (Princeton: Princeton University Press, 2006).

44. Nils Roll-Hansen, *The Lysenko Effect: The Politics of Science* (Amherst, NY: Humanity Books, 2005).

45. For a thorough discussion of the politics of the label "Lysenkoism," see William deJong-Lambert and Nikolai Krementsov, "On Labels and Issues: The Lysenko Controversy and the Cold War," *Journal of the History of Biology* 45 (3) (2012): 373–388, and especially Audra J. Wolfe, "The Cold War Context of the Golden Jubilee, or, Why We Think of Mendel as the Father of Genetics," *Journal of the History of Biology* 45 (3) (2012): 389–414. Earlier materials include David Joravsky, *The Lysenko Affair* (Chicago: University of Chicago Press, 1970), and Valery N. Soyfer, *Lysenko and the Tragedy of Soviet Science* (New Brunswick, NJ: Rutgers University Press, 1994).

46. Gerovitch, *From Newspeak to Cyberspeak*, 547–548.

47. Ibid., 120.

48. Ibid., 126.

49. Mikhail G. Iaroshevskii, "Semanticheskii idealizm: filosofiia imperialisticheskoi reaktsii," in *Protiv filosofii oruzhenostsev amerikano-angliiskogo imperializma*, ed. T. Oizerman and P. Trofimov (Moscow: Nauka, 1951), 100, quoted in Gerovitch, *From Newspeak to Cyberspeak*, 119–121.

50. Gerovitch, *From Newspeak to Cyberspeak*, 119–121.

51. Published under the pseudonym "Materialist," *Vopròsy Filisofii* 5 (1953): 210–219.

52. Gerovitch, *From Newspeak to Cyberspeak*, 124–126. See also *Ocherki istorii informatiki v Russii*, ed. D. Pospelov and Ya. Fet (Novosibirsk: Nauchnyi Tsentr Publikatsii RAS, 1998).

53. Mark M. Rosenthal and Pavel F. Iudin, eds., *Kratkiĭ filosofskiĭ slovar'*, 4th ed. (Moscow: Gospolitizdat, 1954), 236–237; also quoted in Masani, *Norbert Wiener*, 261.

54. Gerovitch, *From Newspeak to Cyberspeak*, 119.

55. Ilia B. Novik, "Normal'naia Lzhnauka" ["A Normal Pseudoscience"], *Voprosy istorii estestvoznaniia i tekhniki (Questions of History of Natural Science and Technology)* 4 (4) (1990), quoted in Gerovitch, *From Newspeak to Cyberspeak*, 103.

56. For more reading on Soviet science, see note 43. On Vygotsky in particular, see Alex Kozulin, *Vygotsky's Psychology: A Biography of Ideas* (Cambridge: Harvard University Press, 1990).

57. Sergei N. Khrushchev, *Nikita Khrushchev and the Creation of a Superpower* (University Park: Pennsylvania State University Press, 2000), 94–96, 163–173.

58. George Paloczi-Horvath, *Khrushchev: The Making of a Dictator* (New York: Little, Brown, 1960), 202.

59. Paloczi-Horvath, *Khrushchev: The Making of a Dictator*, 202.

60. Ibid., 80.

61. David Holloway, "Physics, the State, and Civil Society in the Soviet Union," *Historical Studies in Physical and Biological Sciences* 100 (1) (1999): 173–192; Loren R. Graham, "How Robust Is Science under Stress?," *What Have We Learned about Science and Technology from the Russian Experience?* (Stanford: Stanford University Press, 1998), 52–73.

62. Ivan Pavlov, *Conditioned Reflexes: An Investigation of the Physiological Activity of the Cerebral Cortex*, trans. and ed. G. V. Anrep (London: Oxford University Press, 1927).

63. Josephson, *New Atlantis Revisited*.

64. Stanislav Boguslavski, Henryk Grenievski, and Jerzy Szapiro, "Dialogi o cybernetyce," *Mysl filozoficzna* 4 (14) (1954): 158–212, cited in Günther, "Cybernetics and Dialectical Materialism of Marx and Lenin."

65. Anatoly Kitov, "Chelovek, kotoryi vynes kibernetiku iz sekretnoi biblioteki" ["The Man Who Brought Cybernetics Out of a Secret Library"], interview, *Komp'iuterra* 43 (November 18, 1996): 44–45.

66. Gerovitch, *From Newspeak to Cyberspeak*, 173–175.

67. Ibid., 176–180.

68. Kitov, "Chelovek, kotoryi vynes kibernetiku iz sekretnoi biblioteki," 44–45.

69. Gerovitch, *From Newspeak to Cyberspeak*, 183.

70. Ibid., 173. See also Sergei L. Sobolev, Anatolii I. Kitov, and Aleksei A. Lyapunov, "Osnovnye cherty kibernetiki" ["Basic Features of Cybernetics"], *Voprosy filsofii* 4 (1955): 136.

71. Edwards, *The Closed World*, 175–208, 275–302.

72. Gerovitch, *From Newspeak to Cyberspeak*, 178.

73. Wiener, *Cybernetics*.

74. I discuss the renewability of new media in Peters, "And Lead Us Not into Thinking the New Is New," and Benjamin Peters and Deborah Lubken, "New Media in Crises: Discursive Instability and Emergency Communication," in *The Long History of New Media*, ed. David W. Park et al., 193–209 (New York: Peter Lang, 2011).

75. My thanks to Andriy Ishchenko and an anonymous reviewer for this distinction.

76. Sobolev, Kitov, and Lyapunov, "Osnovnye cherty kibernetiki," 141.

77. Ibid., 141–146.

78. Ibid.

79. Ibid., 147.

80. See Karel Chapek's play *Rossum's Universal Robots*, trans. David Willie (Fairford: Echo Library, 2010).

81. Sobolev et al., "Osnovnye cherty kibernetiki," 148.

82. Ibid., 147.

83. Ibid.

84. Ibid.

85. Ibid.

86. For more on Kolman, see Loren Graham and Jean-Michael Kantor, *Naming Infinity: A True Story of Religious Mysticism and Mathematical Creativity* (Cambridge: Belknap Press of Harvard University Press, 2009).

87. Graham and Kantor, *Naming Infinity*. This fascinating account describes how founding (transfinite) set theorists and religious mystics such as Dmitri Egorov, Pavel Florensky, and Nikolai Luzhin in 1920s Moscow came together around the realization that neither infinity nor God could be defined but both could be named.

88. For more on Lysenko, see deJong-Lambert and Krementsov, "On Labels and Issues." Wolfe, "The Cold War Context of the Golden Jubilee," rethinks the accepted positions against Lysenko laid out in Joravsky, *The Lysenko Affair*, and Soyfer, *Lysenko and the Tragedy of Soviet Science.*

89. Graham and Kantor, *Naming Infinity*, 129.

90. Arnosht (Ernest) Kol'man, *My ne dolzhny byli tak zhit'* [*We Should Not Have Lived That Way*] (New York: Chalidze, 1982), 7, quoted in Graham and Kantor, *Naming Infinity*, 130.

91. Ernest Kolman,"Shto takoe kibernetika?" [What Is Cybernetics?"], *Voprosi Filosophii* (Akademia Nauk CCCR Institut Filosophii, Moscow) 4 (1955): 148–149.

92. Ibid., 149.

93. Wiener briefly studied at Columbia under John Dewey in 1915 and worked as a consultant for a National Defense Research Committee–supported Statistical Research Group based there in 1940.

94. Kolman, "Shto takoe kibernetika?," 150–157.

95. Helmut Dahm, "Zur Konzeption der Kybernetik im dialektischen Materialismus," unpublished manuscript, 25, quoted in Günther, "Cybernetics and Dialectical Materialism of Marx and Lenin," 317–332.

96. David Holloway, for example, writes that "the hostile image of capitalist society, which had played an important part in the early attacks on cybernetics, was now turned to its defense," in "Innovation in Science: The Case of Cybernetics in the Soviet Union," *Science Studies* 4 (1974): 316.

97. Gerovitch, *From Newspeak to Cyberspeak*, 180.

98. Ibid.

99. Sobolev, Kitov, and Lyapunov, "Osnovnye cherty kibernetiki," quoted in Gerovitch, *From Newspeak to Cyberspeak*, 180.

100. Erickson et al., *How Reason Almost Lost Its Mind*, 272.

101. Galison, "The Ontology of the Enemy," 228–266. Peter Galison says that "the enemy as human-machine black box becomes us as human-machine black box." In Sina Najafi and Peter Galison, "The Ontology of the Enemy: An Interview with Peter Galison," *Cabinet* 12 (2003), accessed April 10, 2015, http://cabinetmagazine.org/issues/12/najafi2.php.

102. John A. Armstrong, "Sources of Administrative Behavior: Some Soviet and Western European Comparisons," *American Political Science Review* 59 (3) (1965): 643–655.

103. Gerovitch, "The Cybernetics Scare and the Origins of the Internet," 32–38.

104. D. G. Malcolm, "Review of *Cybernetics at [sic] Service of Communism*, vol. 1," *Operations Research* 11 (1963): 1012.

105. Conway and Siegelman, *Dark Hero of the Information Age*, 316.

106. CIA Intelligence Memorandum, No. 0757/64, "The Meaning of Cybernetics in the USSR," February 26, 1964, 2, also partially quoted in Flo Conway and Jim Siegelman, *Dark Hero of the Information Age* (New York: Basic, 316).

107. CIA Intelligence Memorandum, No. 0757/64, "The Meaning of Cybernetics in the USSR," 3.

108. Ibid., 3.

109. Ibid., 3. See also Gerovitch, "The Cybernetics Scare and the Origins of the Internet," 35.

110. Gerovitch, "The Cybernetics Scare and the Origins of the Internet."

111. Gerovitch, "InterNyet," 340.

112. Gerovitch, *From Newspeak to Cyberspeak*, 249–251.

113. Yu. Kapitonova and A. A. Letichevsky, *Paradigmi i idei akademika V. M. Glushkova* (Kiev: Naukova Dumka, 2003), 296.

114. Igor' A. Poletaev, "O matematicheskom modelirovnanii," *Problemy kibernetiki* 27 (1973): 147.

115. Igor' A. Poletaev, "K opredeleniiu poniatiia 'informatsiia,'" in *Issledovanniia po kibernetike*, ed. A. Lyapunov (Moscow: Sovetskoe radio 1970), 212.

116. Gerovitch, *From Newspeak to Cyberspeak*, 216.

117. Simon Kassel, *Soviet Cybernetics Research: A Preliminary Study of Organizations and Personalities* (Santa Monica: RAND, 1971), v.

118. See the titles of Wiener's *Cybernetics, or Control and Communication in the Animal and the Machine* and his *The Human Use of Human Beings: Cybernetics and Society* (Cambridge: MIT Press, 1950).

119. Gerovitch, *From Newspeak to Cyberspeak*, 208.

120. Ibid., 209–210.

121. Ibid., 210.

122. Pospelov and Fet, *Ocherki istorii informatiki v Rossii*.

123. Conway and Siegelman, *Dark Hero*, 316.

124. Norbert Wiener, "Obschestvo i nauka," *Voprosi Filosofiii* 7 (1961): 49–52.

125. Dirk Jan Struik, "Norbert Wiener: Colleague and Friend," *American Dialog* 3 (1) (1966): 34–37.

126. Bonnie Honig, *Democracy and the Foreigner* (Princeton: Princeton University Press, 2003).

127. On the one hundred twentieth anniversary of his birth and the fiftieth anniversary of his death, the IEEE held a medium-sized conference in Boston on June 24–26, 2014, titled Norbert Wiener in the Twenty-first Century, including a gathering of biographers, former students of his, and rising scholars interested in his life and work.

128. Conway and Siegelman, *Dark Hero*, 314–316. See also Peters, "Toward a Genealogy of Cold War Communication Science."

129. Gerovitch, *From Newspeak to Cyberspeak*, 154–155, 301, passim.

130. James W. Carey with John J. Quirk, "The Mythos of the Electronic Revolution," in *Communication and Culture: Essays on Media and Society*, 113–141 (New York: Unwin Hyman, 1989).

131. For more on feedback in the Western political tradition, see Otto Mayr, *Authority, Liberty, and Automatic Machinery in Early Modern Europe* (Baltimore: Johns Hopkins University Press, 1989), 144; see also Bernard Geoghegan, "The Cybernetic Apparatus: Media, Liberalism, and the Reform of the Human Sciences," Ph.D. diss., Northwestern University, Chicago, 2012.

132. Gerovitch, *From Newspeak to Cyberspeak*, 122.

133. Graham, *Lonely Ideas*, 1–4. For contrasting portraits of the local contingencies and practices that animate laboratory work, see Galison, *Image and Logic*, and Latour, *Science in Action*.

134. Dima Adamsky, *The Culture of Military Innovation: The Impact of Cultural Factors on the Revolution in Military Affairs in Russia, the US, and Israel* (Stanford: Stanford University Press, 2010), 37, 24–57.

135. Medina notes that "Beer was well aware of the Soviet approach to cybernetic management, and he viewed it with open contempt." Eden Medina, *Cybernetic Revolutionaries: Technology and Politics in Allende's Chile* (Cambridge: MIT Press, 2011), 63.

136. Herbert Spencer, *The Principles of Sociology*, 3rd ed. (New York: Westminster, [1876] 1896), 460–462, 478–545.

137. Charles Horton Cooley, *Sociological Theory and Social Research: Selected Papers of Charles Horton Cooley* (New York: Henry Holt, 1930), 6.

138. O'Shea, *The Brain*, 1.

Chapter 2: Economic Cybernetics and Its Limits

1. Regarding the mutual embeddedness of practice and theory on which this analysis of Soviet economic problems rests, I take for granted (more or less following John Dewey and other early pragmatists) that the two cannot be separated. Without practice, theory is a mere abstraction, a desiccation of thought; without theory, practice is purposeless action. I understand theory as a form of practice, however subdued and meditative its rootedness in modern society may be, and practice as an expression of mental purpose, an exercising of theory in a world that knows only action. C. S. Peirce put the point thus: "Consider what effects, which might conceivably have practical bearings, we conceive the object of our conception to have. Then, our conception of those effects is the whole of our conception of the object." In other words, to consider an object is to conceive of its practical effects. To conceive, or theorize, an object is, for the early pragmatists, also to understand the full set of its practices and implications. With this in mind, the analysis of organizations and economics that follows assumes that theoretical and practical judgments must be reconcilable. For more, see John Dewey, *Logic: The Theory of Inquiry*, in *The Essential Dewey*, vol. 2 (Bloomington: Indiana University Press, 1999), 169–179; see also Charles Sanders Peirce, "How to Make Our Ideas Clear," in *The Essential Peirce*, vol. 1 (Bloomington: Indiana University Press, 1992–1999), 132.

2. Before 1928, the Soviet Union was an indicative economy, not a command economy, meaning that the state set economic quotas but did not compel them. Richard E. Ericson, "Command Economy," *The New Palgrave Dictionary of Economics*, 2nd ed., ed. Steven N. Durlauf and Lawrence E. Blume (New York: Palgrave, 2008).

3. Engels wrote to Karl Kautsky in Vienna: "In any case, it will be for those people to decide if, when and what they want to do about it, and what means to employ. I don't feel qualified to offer them any advice or counsel in this matter. They will presumably be at least as clever as we are." Friedrich Engels to Karl Kautsky in Vienna, from Karl Marx and Friedrich Engels, *Karl Marx and Frederick Engels: Selected Correspondence* (Moscow: Progress, 1975), accessed July 25, 2013, http://www.marxists.org/archive/marx/works/1881/letters/81_02_01.htm.

4. Stephen F. Cohen, *Bukharin and the Bolshevik Revolution: A Political Biography, 1888–1938* (New York: Oxford University Press, 1971), 93.

5. Much of the vast literature on the Soviet command economy is dated to cold war research concerns. The part that was consulted (and sometimes critiqued) in this work includes Mark Beissinger, *Scientific Management, Socialist Discipline, and Soviet Power* (Cambridge: Harvard University Press, 1988); Peter Blau, *Bureaucracy in Modern Society* (New York: Random House, 1956); Michael Ellman, *Planning Problems in the USSR: The Contributions of Mathematical Economics to Their Solution, 1960–1971* (New York: Cambridge University Press, 1973); Michael Ellman, *Socialist Planning* (New York: Cambridge, 1978); Paul R. Gregory, *The Political Economy of Stalinism: Evidence*

from the Soviet Secret Archives (New York: Cambridge University Press, 2003); Paul R. Gregory, *Restructuring the Soviet Economic Bureaucracy* (New York: Cambridge University Press, 1990); Gregory Grossman, ed., *Studies in the Second Economy of Communist Countries: A Bibliography* (Berkeley: University of California Press, 1988); Gregory Grossman, "Notes for a Theory of the Command Economy," *Soviet Studies* 15 (2) (1963): 101–123; Gregory Grossman, "The 'Second Economy' of the USSR," *Problems of Communism* 26 (5) (1977): 25–40; János Kornai, *The Socialist System*; Alena V. Ledeneva, *Russia's Economy of Favors: Blat, Networking, and Informal Exchange.* New York: Cambridge University Press, 1998; Alec Nove, *An Economic History of the USSR, 1917–1991*, 3rd ed. (New York: Penguin, 1992); Elena Osokina, *Our Daily Bread: Socialist Distribution and the Art of Survival in Stalin's Russia, 1927–1941* (New York: Routledge, 2003); and Alejandro Portes, Manuel Castells, and Lauren A. Benton, eds., *The Informal Economy: Studies in Advanced and Less Developed Countries* (Baltimore: Johns Hopkins University Press, 1989).

6. George M. Armstrong Jr., *The Soviet Law of Property* (The Hague: Martinus Nijhoff, 1983). See also John N. Hazard, *Communists and Their Law: A Search for the Common Core of the Legal Systems of the Marxian Socialist States* (Chicago: University of Chicago, 1969), 171–223; and Karl Marx and Friedrich Engel, "Manifesto of the Communist Party" (1848) in Robert C. Tucker, ed., *The Marx-Engels Reader, 2nd Edition* (New York: W. W. Norton, 1978): "the theory of the Communists may be summed up in the single sentence: abolition of private property."

7. David Dyker, *Restructuring the Soviet Economy* (New York: Routledge, 1991), 7.

8. Mark Harrison, "Soviet Economic Growth since 1928: The Alternative Statistics of G. I. Khanin," *Europe-Asis Studies* 45 (1) (1993): 141–167.

9. Estimates of the number of victims range from roughly 7 million to 14 million. Robert Conquest, *The Harvest of Sorrow* (New York: Oxford University Press, 1986); Timothy Snyder, *Bloodlands: Europe between Hitler and Stalin* (New York: Basic Books, 2012); Miron Dolot, *Execution by Hunger: The Hidden Holocaust* (New York: Norton, 2011).

10. David C. Engerman, *Modernization from the Other Shore: American Intellectuals and the Romance of Russian Development* (Cambridge: Harvard University Press, 2003).

11. Anders Aslund, *How Capitalism Was Built: The Transformation of Central and Eastern Europe* (New York: Cambridge University Press, 2007), 75; Noel E. Firth and James H. Noren, *Soviet Defense Spending: A History of CIA Estimates, 1950–1990* (College Station: Texas A&M University Press).

12. Francis Spufford, in his delightful novel *Red Plenty*, fabricates a relatable incident in which a brake failure sends a tractor hurtling through the wall of a crucial factory and thereby disrupts the production of a specific large piece of machinery for months. The disruption sends a ripple of delays and costs across the national indus-

tries that depend on the factory for the machine that it produces. Francis Spufford, *Red Plenty* (Minneapolis, MN: Graywolf Press, 2012).

13. I. Borovitski, editorial, *Pravda*, October 3, 1962.

14. N. Chesnenko, "'Obshchii iazyk' elektronnykh mashin: Problemy kodirovaniia dannykh," *Ekonomiceskaia gazeta* (47) (1973): 10.

15. Gertrude E. Schroeder, "Organizations and Hierarchies: The Perennial Search for Solutions," in *Reorganization and Reform in the Soviet Economy*, ed. Susan J. Linz and William Moskoff (New York: Sharpe, 1988), 6.

16. Leon Smolinski, "What Next in Soviet Planning?," *Foreign Affairs* 42 (3) (1964): 603–613.

17. Aleksei Kuteinikov, "Pervie proekti avtomatizatsii upravleniya sovetskoi planovoi ekonomikoi v kontse 1950-x I nachale 1960-x gg.—'elektronnyi sotsializm'?," *Ekonomicheskaya istoriya* (Moscow: Trudi istoricheskogo faku'teta MGU) 15 (2011): 126.

18. Castells, *End of the Millennium*, 17, see also 5–68.

19. Alex Galloway, *Protocol: How Control Exists after Decentralization* (Cambridge: MIT Press, 2004), 2–28, 240–247.

20. E. G. Liberman, "Plans, Profits and Bonuses," *Pravda*, September 9, 1962, quoted in *The Liberman Discussion: A New Phase in Soviet Economic Thought*, ed. M. E. Sharpe (White Plains, NY: International Arts and Science Press, 1965), 000–000.

21. John Marangos, *Consistency and Viability of Socialist Economic Systems* (New York: Palgrave Macmillan, 2013), esp. chapter 5.

22. David Alexander Lax, *Libermanism and the Kosygin Reform* (Charlottesville: University of Virginia Press, 1991).

23. William Taubman, Sergei Khrushchev, and Abbott Gleason, *Nikita Khrushchev* (New Haven: Yale University Press, 2000), 153–154.

24. Karl W. Ryavec, *Russian Bureaucracy: Power and Pathology* (New York: Rowman & Littlefield, 2005), 227–230.

25. Gottfried Liebniz, "The Art of Discovery" (1685), in *Leibniz: Selections*, ed. Philip P. Wiener (New York: Charles Scribner's Sons, 1951), 50–58.

26. Kantorovich and von Neumann were born to middle-class Jewish families in eastern Europe. Roy Gardner, in "L. V. Kantorovich: The Price Implications of Optimal Planning," in *Socialism and the Market: Mechanism Design Theory and the Allocation of Resources*, ed. Peter J. Boettke, 638–648 (New York: Routledge, 2000). Few have satisfactorily described the forces behind the phenomenal scientific output of the generation born between 1890 and 1930 to a tiny Jewish middle class in Hungary.

Members of this group include mathematician and founding computer scientist and game theorist John von Neumann; pan-prolific mathematician Paul Erdős; Nobel laureate and founder of holography Dennis Gabor; Nobel laureate and physicist Eugene Wigner; early supersonic aerospace engineer Theodore von Kármán; discoverer of the linear accelerator, the electron microscope, and nuclear chain reaction Leo Szilard; the primary force behind the hydrogen bomb Edward Teller; codeveloper of BASIC computer programming John George Kemeny; historian Oszkar Jaszi; philosopher Georg Lukacs, economist and philosopher Karl Polanyi; author Arthur Koesler; and composer Bela Bartok. Gabor Pallo, "The Hungarian Phenomenon in Israeli Science," *Bulletin of the History of Chemistry* 25 (1) (2000): 35–42.

27. Independent of Kantorovich, Von Neumann and George Dantzig developed similar methods in the United States after the war.

28. To add some numbers to the basic problem: assume that a square meter of potatoes costs two rubles to grow and sells at six rubles and that a square meter of wheat costs three rubles and sells at seven rubles. Given 100 square meters, the linear programmer might ask, "What proportion of potatoes and wheat will maximize revenue?" In practice, programmers struggled to address massively more complicated programs, factoring into their matrices and algorithms the constraints, costs, and effects of dozens or hundreds of variables from pesticides, fertilizer, and soil degradation.

29. Iosif V. Stalin, *Voprosy leninizma*, 11th ed. (Moscow: Gospolitizdat, 1951), 326.

30. Abraham S. Becker, "Input-Output and Soviet Planning: A Survey of Recent Developments," paper prepared for the United States Air Force Project RAND, Memorandum, RM 3523-PR, March 1963, accessed July 18, 2013, http://www.dtic.mil/cgi-bin/GetTRDoc?AD=AD0401490.

31. V. S. Nemchinov, *O dalneishem sovershenstvovanii planirovaniya i upravleniya narodnym khozyaistvom*. Moscow: Ekonomika, 1964: 1–74. V. S. Nemchinov, "Sotsialisticheskoe khozyaistvovanie i planirovanie proizvodstva," *Kommunist* (1964): 5.

32. A. Birman, "Neotvratimost," *Zvezda* 5 (1978): 1–5.

33. Boris Nikolaevich Malinovsky, *Pioneers of Soviet Computing*, ed. Anne Fitzpatrick, trans. Emmanuel Aronie, 2010, accessed April 15, 2015, http://www.sigcis.org/files/SIGCISMC2010_001.pdf, esp. "Personal Reminisces of Viktor Glushkov," 34–59.

34. Adamsky, *The Culture of Military Innovation*, 26–31; Kapitonova and Letichevsky, *Paradigmi i idei akademika V.M. Glushkova*, 164.

35. Stark, *The Sense of Dissonance*, 1–34, see also 35–51, 54–80.

36. Spufford, *Red Plenty*, 208–209.

37. Kornai, *The Socialist System*, 121.

38. Ibid., 121.

39. Ibid., 122.

40. Ibid., 122–123.

41. Castells, *End of the Millennium*, 24.

42. For a basic review of *tolkachy* and other informal mechanisms in the economy, see Mark Beissinger, *Scientific Management, Socialist Discipline, and Soviet Power* (Cambridge: Harvard University Press, 1988); Ledeneva, *Russia's Economy of Favors*; and Alena V. Ledeneva, *Can Russia Modernize? Sistema, Power Networks and Informal Governance* (New York: Cambridge University Press, 2013).

43. Byung-Yeon Kim, "Informal Economy Activities of Soviet Households: Size and Dynamics," *Journal of Comparative Economics* 31 (3) (2003): 532–551.

44. Kim, "Informal Economy Activities of Soviet Households," 532–535; Simon Johnson, Daniel Kaufmann, and Andrei Shleifer, "The Unofficial Economy in Transition," *Brookings Papers on Economic Activity* 2 (1997): 159–221.

45. Ledeneva, *Russia's Economy of Favors*, 12.

46. Zbigniew K. Brzezinski, *The Soviet Block: Unity and Conflict*, rev. ed. (New York: Praeger, 1960), 116, see also 115–124.

47. From *Elet es Tudomany*, December 24, 1952, and *Rude Pravo*, December 21, 1952, quoted in Brzezinski, *The Soviet Block*, 114.

48. David Granick, *Management of the Industrial Firm in the USSR: A Study in Soviet Economic Planning* (New York: Columbia University Press, 1955), 229.

49. Gregory, *Restructuring the Soviet Economic Bureaucracy*, 173. On the sticking power of informal relations in other socially networked economies, see Mark Granovetter, "The Strength of Weak Ties," *American Journal of Sociology* 78 (6) (1973): 1360–1380.

50. Gertrude Schroeder, "The Soviet Economy on a Treadmill of Reforms," *Soviet Economy in a Time of Change*, U.S. Congress Joint Economic Committee (Washington, DC: USGPO, 1979).

51. Castells, *The End of the Millennium*, 24.

52. Loren R. Graham, *The Ghost of the Executed Engineer: Technology and the Fall of the Soviet Union* (Cambridge: Harvard University Press, 1993), 73.

53. Thorsten Veblen, *The Engineers and the Price System* (New York: Huebsch, 1921).

54. Castells, *The End of the Millennium*, 30.

55. Erickson et al., *How Reason Almost Lost Its Mind*, 81–106.

Chapter 3: From Network to Patchwork

1. V. A. Kitov, E. N. Filinov, and L. G. Chernyak, "Anatoly Ivanovich Kitov," accessed May 19, 2010, http://www.computer-museum.ru/galglory/kitov.htm; and Vladimir A. Kitov and Valery V. Shilov. "Anatoly Kitov: Pioneer of Russian Informatics," in *History of Computing: Learning from the Past*, vol. 325, ed. Arthur Tatnall (New York: Springer, 2010), 80–88.

2. Slava Gerovitch, *From Newspeak to Cyberspeak* (Cambridge: MIT Press, 2002), 138–139.

3. Richard J. Samuels, *"Rich Nation, Strong Army": National Security and the Technological Transformation of Japan* (Ithaca: Cornell University Press, 1994), 1–32.

4. See Edwards, *The Closed World*, 75–115, esp. 99–100; see also Thomas Hughes's *Rescuing Prometheus: Four Monumental Projects That Changed the Modern World* (New York: Vintage, 2000), esp. chap. 2 on SAGE and chap. 4 on ARPANET.

5. Kitov, Filinov, and Chernyak, "Anatoly Ivanovich Kitov."

6. Gerovitch, "InterNyet," 338–339. See also Theodore Shabad, "Khrushchev Says Missile Can 'Hit a Fly' in Space," *New York Times*, July 17, 1962. Marshal Rodion Malinovsky, the minister of defense, made a similar claim more carefully several months earlier: "the problem of destroying ballistic missiles in flight has been successfully solved" as reported in an unnamed article in *Pravda*, October 25, 1961. For imaginatively named radar networks, see Ashton B. Carter and David N. Schwartz, *Ballistic Missile Defense* (Washington, DC: Brookings Institution Press, 1984), 197–198.

7. Recent technology commentators and scholars have enthused about the analog update of "cognitive surplus" that can be made available over collaborative peer-based computer networks. See, for example, Clay Shirky, *Cognitive Surplus: Creativity and Generosity in a Connected Age* (New York: Penguin Books, 2010), and Yochai Benkler, *The Wealth of Networks: How Social Production Transforms Markets and Freedom* (New Haven: Yale University Press, 2006), 35–132.

8. Anatoly Kitov, *Electronnie tsifrovie mashini* [*Electronic Ciphered Machines*] (Moscow: Radioeletronika Nauka, 1956).

9. Charles Eames and Ray Eames, *A Computer Perspective: Background to the Computer Age* (Cambridge: Harvard University Press, 1973), 64, 96–97.

10. Marc Raeff, *The Well-Ordered Police State: Social and Institutional Change through Law in the Germanies and Russia, 1600–1800* (New Haven: Yale University Press, 1983); Jacob Soll, *The Information Master: Jean-Baptiste Colbert's Secret State Intelligence System* (Ann Arbor: University of Michigan Press, 2009).

11. Eduard A. Meerovich, "Obsuzhdenie doklada professor A. A. Liapunova 'Ob ispol'zovanii matematicheskikh mashin v logicheskikh tseliakh" (1954), in *Ocherki istorii informatiki*, ed. D. Pospelov and Ya. Fet (Novosibirsk: Nauchnyi Tsentr Publikatsii RAS, 1998), accessed March 20, 2010, http://ssd.sscc.ru/PaCT/history/early. html,75.

12. Isaak C. Bruk, "Elektronnie vyichislitel'nie mashinyi—na sluzhbu narodnomu khozyaistvu," *Kommunist* 7 (1957): 127.

13. A. I. Kitov, "Pis'mo zamestitelya nachal'nika VTs Minoboroni SSSR A.U. Kitova v TsK KPSS N.S. Khrushchyovu ot 7 Anvarya 1959 goda," Politekhnicheskii museu RF, fond "Kitov Anatolii Ivanovich," f. 228, edinitsa khraneniya KP27189/20.

14. Anatoly Kitov, "Rol' akademika A. I. Berga v razvitii vyichislitel'noi tekhniki I avtomatizirovannikh system upravleniya" ("The Role of Academician A. I. Berg in the Development of Computational Technology and Automated Management Systems"), in *Put' v bol'shuyu nauku: Akademik Aksel' Berg [Pathway to Big Science: Academician Aksel Berg]* (Moscow: Nauka, 1988), accessed March 20, 2010, http://www .computer-museum.ru/galglory/berg3.htm.

15. Anatoly Kitov, "Letter to Khrushchev," January 7, 1959, 2, Politechnical Museum of the Russian Federation, Collection Kitov, Anatoly Ivanovich, file 228, unit of storage KP27189/20.

16. Ibid., 1–2.

17. Anatoly Kitov, "Chelovek, kotoryi vynes kibernetiku iz sekretnoi biblioteki" ["The Man Who Brought Cybernetics out of a Secret Library"], *Komp 'iuterra* 18 (43) (1996): 44–45.

18. Ibid.

19. Aleksei Kuteinikov, "Pervie proekti avtomatizatsii upravleniya sovetskoi planovoi ekonomikoi v kontse 1950-x I nachale 1960-x gg.—'elektronnyi sotsializm'?" *Ekonomicheskaya istoriya*, 124–138 (Moscow: Trudi istoricheskogo faku'teta MGU 15, 2011), 109.

20. For a hagiographic biographical blurb (in Russian) on Konstantin Konostantinovich Rokossowski, see his memorial site, accessed July 25, 2012, http://www .rokossowski.com/bio.htm.

21. Lewis Mumford, *Technics and Civilization* (New York: Harvest Book, 1934).

22. Kitov, "Chelovek, kotoryi vynes kibernetiku iz sekretnoi biblioteki."

23. For more on Soviet military, see Roger R. Reese, *The Soviet Military Experience* (New York: NP, 2000), and William E. Odom, *The Collapse of the Soviet Military* (New Haven: Yale University Press, 1998).

24. Kitov, ""Chelovek, kotoryi vynes kibernetiku."

25. A sample of Kitov's publications relevant to the EASU can be found here, accessed July 25, 2013: http://www.kitov-anatoly.ru/naucnye-trudy/perecen-osnovnyh-naucnyh-trudov.

26. Berg, Kitov, and Lypunov, "O vozmozhnostyakh avotmatizatsii upravleniya narodniym kozyaistvom."

27. Postanovlenie TsK KPCC I Soveta Ministrov SSSR, "Ob uluchshenii rukovodstva vnedreniem bychislitel'noi tekhniki i avtomatizirovannikh system upravleniya v narodnoe khozyaistvo," May 21, 1963, Gosudarstvenni archive (GA RF): f. 5446, o. 106, d. 1324, l. 160–172. This document is published in full for the first time in Aleksei Viktorovich Kuteinikov, "Proekt Obshchegosudarstvennoi avtomatizirovan-noi sistemi upravleniya sovetskoi ekonomikoi (OGAS) i problem ego realizatsii v 1960–1980-x gg."

28. For example, one year earlier, Boris Pasternak was ridiculed in literary circles for being awarded and then declining the Nobel Prize in literature. It is another case of being punished for appealing too successfully to unapproved authorities. For more on the campaign against Pasternak and others, see Solomon Volkov, *The Magical Chorus: A History of Russian Culture from Tolstoy to Solzhenitsyn* (New York: Knopf, 2008), 195–196.

29. Boris Nikolaevich Malinovskii, *Istoriaa vychislitel'noi tekhniki* (Kiev: Gorobets, 2007), 197–207, accessed April 15, 2015, http://lib.ru/MEMUARY/MALINOWSKIJ.htm.

30. J.C.R. Licklider, "Man-Machine Symbiosis," *IRE Transactions on Human Factors in Electronics* HFE-1 (March 1960): 4–11, see section 5.1, "Speed Mismatch between Men and Computers," accessed July 25, 2013, http://groups.csail.mit.edu/medg/people/psz/Licklider.html.

31. This phrase comes from a CIA declassified document that the author received via a Freedom of Information Act request. For the phrase "unified information net-work," see CIA declassified documents by John J. Ford: A. "The Meaning of Cyber-netics in the USSR," Intelligence Memorandum No. 0757/64, February 26, 1964, 1–10, esp. 1. See also B. "The Cybernetic Approach to Education in the USSR," Scien-tific Intelligence Memorandum No. 464693, May 25, 1964; C1. "The Soviet Applica-tions of Cybernetics in Medicine: 1. Medical Diagnosis," Scientific and Technical Intelligence Report No. 464692, September 15, 1966; C2. "2. Artificial Limbs," Scien-tific and Technical Intelligence Report No. 464691, May 10, 1967; D. "Major Devel-opments in the SovBloc Cybernetics Programs in 1965," Scientific and Technical Intelligence Report No. 464694, October 3, 1966, 1–33.

32. Robert A. Divine, *The Sputnik Challenge: Eisenhower's Response to the Soviet Satellite* (New York: Oxford University Press, 1993), 104, 111–112, 146–155.

33. Janet Abbate, *Inventing the Internet* (Cambridge: MIT Press, 1999); see also Peter H. Salus, ed. *The ARPANET Sourcebook* (Charlottesville, VA: Peer-to-Peer Communications LLC, 2008).

34. Another "humor-neutic" reading might have God speaking Hebrew through the wires (*lo* means "no" or "not" in modern Hebrew).

35. Abbate, *Inventing the Internet*, 75–77.

36. Audra J. Wolfe, *Competing with the Soviets: Science, Technology, and the State in Cold War America* (Baltimore: Johns Hopkins University Press, 2013), 49–50, ibid., esp. chaps. 2 and 3; Stuart W. Leslie, *The Cold War and American Science: The Military-Industrial-Academic Complex* (New York: Columbia University Press, 1993), 203–231.

37. Kristie Mackrasis, *Seduced by Secrets: Inside the Stasi's Spy-Tech World* (New York: Cambridge University Press, 2014), 23, 133, 139, esp. 112–140.

38. Judy O'Neill, "Interview with Paul Baran," Charles Babbage Institute, OH 182, March 5, 1990, Menlo Park, CA, accessed April 15, 2015, http://www.gtnoise.net/classes/cs7001/fall_2008/readings/baran-int.pdf.

39. Ibid.; see also Stewart Brand, "Founding Father," *Wired* 9 (3) (1991), accessed April 15, 2015, http://archive.wired.com/wired/archive/9.03/baran_pr.html.

40. Brand, "Founding Father."

41. Ibid.

42. Bradley Voytek, "Are There Really as Many Neurons in the Human Brain as Stars in the Milky Way?," *Nature* (Scitable blog, May 20, 2013), accessed April 15, 2015, http://www.nature.com/scitable/blog/brain-metrics/are_there_really_as_many.

43. Katie Hafner and Matthew Lyon, *Where the Wizards Stay up Late* (New York: Simon & Schuster, 1996), 64.

44. James Carey describes the process of communication as "models of and for reality that make the world apprehensible" in "A Cultural Approach to Communication," *Communication as Culture: Media and Society* (New York: Unwin Hyman, 1989), 32.

45. Aleksandr Kharkevich, "Informatsia i tekhnika" ["Information and Technology"], *Kommunist* 17 (1962): 94.

46. Ibid. For an example of his earlier and largely technocratic information theory work, see Aleksandr A. Kharkevich, "Basic Features of a General Theory of Communication," *Radiotekhnika* [Radio Engineering] 9 (5) (1954). For the CIA document, see Conway and Siegelman, *Dark Hero of the Information Age*, n. 318.

47. Shannon, "A Mathematical Theory of Communication."

48. Ibid., 102.

49. Kharkevich, "Informatsia i teckhnika," 102.

50. Ibid., 94.

51. For the first public formulation of Moore's law, see Gordon E. Moore, "Cramming More Components onto Integrated Circuits," *Electronics* 38 (8) (1965): 114–117.

52. For the first systematic work to treat knowledge as an economic measure and resource and thus to anticipate the accounting of postindustrial information and service sectors, see Fritz Machlup, *The Production and Distribution of Knowledge in the United States* (Princeton: Princeton University Press, 1962).

53. Kharkevich, "Informatsia i tekhnika," 102.

54. Ibid., 102.

55. Ibid., 103.

56. Ibid., 102.

57. N. I. Kovalev, "Doklad o rabote i perspektivakh razvitiya VTs pri Gosekonomsovete" ["Report about the Work and Perspectives of the Development of Information Technology in the Gosekonomsovet"], July 23, 1962, Rossiiskii gosudarstvenniyi arkhiv ekonomiki (RGAE) [Russian State Archive of Economics], Moscow, f. 9480, o. 7, d. 466, l. 77–97, quoted in Kuteinikov, "Pervie proekti," 134 n. 3.

58. Ibid., quoted in Kuteinikov, "Pervie proektyi," 134–135.

59. For more on the cultural complications of automation as a Soviet concept, see Slava Gerovitch, "Human-Machine Issues in the Soviet Space Program," *Critical Issues in the History of Spaceflight*, ed. Steven J. Dick and Roger D. Launius, 107–140 (Washington, DC: NASA History Division, 2006).

60. Rasmus Kleis Nielsen, "Democracy," in *Digital Keywords: A Vocabulary for Information Society and Culture*, ed. Benjamin Peters (Princeton: Princeton University Press, under review), accessed April 15, 2015, http://culturedigitally.org/2014/05/democracy-draft-digitalkeywords. See also John Keane, *The Life and Death of Democracy* (New York: Simon & Schuster, 2009).

Chapter 4: Staging the OGAS, 1962 to 1969

1. John Lewis Gaddis, *The Cold War: A New History* (New York: Penguin Press, 2005), 78.

2. V. Glushkov, "Kibernetika, progress, budushchee," *Literaturnaya Gazeta*, September 25, 1962, 1–3.

3. "Voprosi Strukturi, Organizatsii i sozdaniya edinoi gosudarstvennoi seti vyichislitel'nikh tsentrov EGSVTs," *Rossiiskii gosudarstvennyi arkhiv ekonomiki* (RGAE), f. 9480, o. 7, d. 1227, l. 82–102, reproduced in full in the appendix to Aleksei Viktorovich Kuteinikov, "Proekt Obshchegosudarsvetnnoyi avtomatizirovannoi sistemi upravleniya sovetskoi ekonomikoi (OGAS) i problem ego realizatsii v 1960–1980-x gg," Ph.D. diss., Moscow State University, 2011.

4. Aleksei Viktorovich Kuteinikov, "Proekt avtomatizirovannoi sistemi upravleniya sovetskoi ekonomikoi (OGAS) i problem ego realizatsii v 1960–1980" ["The project of all-state automated system of management of Soviet economics (OGAS) and the problem of its realization in 1960–1980"], Ph.D. diss., Moscow State University, 2011.

5. Vincent Mosco, *To the Cloud: Big Data in a Turbulent World* (New York: Paradigm Publishers, 2014).

6. Gerovitch, *From Newspeak to Cyberspeak*, 283.

7. Gerovitch, "InterNyet," 341; Medina, *Cybernetic Revolutionaries*, 75.

8. Viktor Glushkov, uncollected archives in the closet of the main office, Institute of Cybernetics, room 804, Kiev, Ukraine.

9. Kuteinikov, "Proekt Obshchegosudarstrnnoi."

10. Author interview with Vera Glushkov, May 14, 2012.

11. Malinovsky, *Pioneers in Soviet Computing*, 31–34.

12. Kapitonova and Letichevsky, *Paradigmi i idei akademika V. M. Glushkova*, 225–232.

13. Ibid.

14. Ibid., 142–144.

15. Malinovsky, *Istoriya vyicheslitel'noi tekhniki*, 92–93.

16. Kapitonova and Letichevski, *Paradigmi i idei akademika V. M. Gluschkova*, 316–317, see also 296–317.

17. Ibid.

18. From Gluskov's unpublished memoirs "Vopreki Avtoritetam" ["Despite the Authorities"], accessed April 15, 2015, http://lib.ru/MEMUARY/MALINOWSKIJ/5.htm.

19. Unsourced quote on the title screen accessed April 15, 2015, http://ogas.kiev.ua.

20. Aleksandr Ivanovich Stavchikov, "Romantika pervyikh issledovanii i proektov i ikh protivorechnaya sud'ba" ["Romanticism of Early Research and Projects and Their Contradictory Fate"], unnamed, unpublished history of Central Economic

Mathematical Institute (CEMI), Moscow, read in person and returned May 2008, chap. 2, 17. See CEMI-RAS Archive in bibliography.

21. Anatoly Kitov, "Kibernetika i upravlenie narodnym khoziastvom" ["Cybernetics and the Management of the National Economy"], *Kibernetiku—na sluzhbu Kommunism* (*Cybernetics: In the Service of Communism*), ed. Aksel' Berg, 1 (1961): 207, 216.

22. Eden Medina, *Cybernetic Revolutionaries: Technology and Politics in Allende's Chile* (Cambridge: MIT Press, 2011), 35, 34–39, 75–76.

23. Malinovsky, *Pioneers of Soviet Computing,* including excerpts of Glushkov's unpublished memoirs, "Vopreki avtoritetam" ["Despite the Authorities"], accessed April 15, 2015, http://lib.ru/MEMUARY/MALINOWSKIJ/5.htm.

24. An early mention in English of the "paperless office" can be found in "The Office of the Future," *Business Week* 30 (2387) (1975): 48–70. See also Abigail Sellen and Richard Harper, *The Myth of the Paperless Office* (Cambridge: MIT Press, 2003); Paul A. Marolla et al., "A Million Spiking-Neuron Integrated Circuit with a Scalable Communication Network and Interface," *Science* 8 (345) (2014): 668–673, accessed April 15, 2015, http://www.sciencemag.org/content/345/6197/668.

25. Kapitonova and Letichevsky, *Paradigm i idei akademika V. M. Glushkova*, 18.

26. Ibid., 18.

27. Their friendship eventually became a family relationship. In the 1980s, Kitov's son Vladimir married Glushkov's oldest daughter, Olga, who raised a grandson named Viktor. Glushkov's youngest daughter, Vera, also named Glushkov's granddaughter Viktoria.

28. Author's interview with Boris Nikolaevich Malinovsky, April 7, 2012.

29. The title of Glushkov's last scholarly book was *Fundamentals of Paperless Informatics*. Viktor Glushkov, *Osnovi bezbumazhnoi informatiki* (Moscow: Nauka, 1982), 552.

30. Quoted in Gerovitch, "InterNyet," 345; Viktor Glushkov, *Kibernetika, vychislitel'naia tekhnika, informatika. Izbrannye trudy* (Kiev: Naukova dumka, 1990), 92.

31. Gerovitch, "InterNyet," 342–345.

32. Vladislav Zubok, *Zhivago's Children: The Last Russian Intelligentsia* (Cambridge: Harvard University Press, 2009), 275.

33. At least two politically distinct scholars have made this same basic point forcefully in the last decade: Niall Ferguson, *The Ascent of Money: A Financial History of the World* (New York: Penguin Group, 2008); Graeber, *Debt: The First Five Thousand Years*.

34. Glushkov, "Shto skazhet istoria," 3, accessed April 15, 2015, http://ogas.kiev.ua/history/chto-skazhet-ystoryya.

35. Author's interview with Vera Viktorevna Glushkov, April 30, 2012.

36. Gerovitch, "InterNyet."

37. Boris Nikolaevich Malinovsky, *Vechno Khranit* [*Store Eternally*] (Kiev: Gorobets, 2007), 58.

38. These details are summarized from four documents in the archival materials in Viktor M. Glushkov's personal files, box 18, folder 1, documents 12, 119, 122, and 123 inclusive, at the Archive and Special Collections, National Academy of Sciences of Ukraine, Kiev. Nancy Ries also examines the culture of institutional authorities as a form of moral power in *Russian Talk: Culture and Conversation during Perestroika* (Ithaca: Cornell University Press, 1997), 88–89.

39. See the memoir of the leading participants in Lebedev's team: Sergei Lebedev, Lev Dashevsky, and Ekaterina Shkabara, "Malaya elektronnaya shchyotnyaya mashina" ["Small Electronic Digital Computer"], 1952, accessed April 15, 2015 http://it-history.ru/images/a/af/SALebedev_MESM.pdf; Malinovsky, *Istoria vyichislitel'noi tekhniki v litsakh*, 33–34.

40. See the only known other secondary document on Cybertonia, Vera V. Glushkova and Sergei A. Zhabin, "Virtualnaya strana Kibertonia v Institute Kibernetiki (60–70 gg, XX vek)," in *Ukrainia i svit: gumanitarno-tekhnicheska elita ta sotsialnyi progress: tezi dopov* [*Ukraine and the World: Humanitarian-Technical Elite and Social Progress*], supplementary theses, International Scientific-Theoretical Conference for Students and Graduate Students, April 4–5, 2012 (Kharkiv: NTU Kharkiv, 2012), 81–83.

41. Personal correspondence with Vera Viktorevna Glushkova, February 28, 2012.

42. Public press on Cybertonia includes clippings from "Vechirnii Kiiv," 305 (5624) (December 31, 1962): 3, and "Vechirniy Kiev," 309 (6588) (December 31, 1965): 2–3. In parody of and in the same font as *Vechernii Kiev*, the group also issued its own *Vechernyi Kyber* as the "newspaper of the council of robots" 1 (1) (1966). See also "Podorozh v Krainu Kibertonii" ["Travel to the Country of Cybertonia"], *Kievskii Komsomoltsyi* 1 (1014) (August 1, 1963): 2–3, and A. Voloshin, "Kibertonia-65," *Vechirnii Kiev* (February 16, 1965): 2. All documents are retained in the author's personal archives.

43. Vera Viktorevna Glushkova, "Dorogoi chitatel', dobro pozhalovat' v 'kibertoniyu'!," *Cybertonia* 1 (1) (2012): 2, accessed April 15, 2015, http://miratechgroup.com/sites/default/files/documents/press_about_us/kibertonia_n01-2012.pdf.

44. For references on jazz in the Cold War and the Soviet Union, see S. Frederick Starr, *Red and Hot: The Fate of Jazz in the Soviet Union, 1917–1991* (New York: Oxford

University Press, 1994); Lisa E. Davenport, *Jazz Diplomacy: Promoting America in the Cold War Era* (Jackson: University Press of Mississippi, 2009); and Penny von Eschen, *Satchmo Blows up the World: Jazz Ambassadors Play the Cold War* (Cambridge: Harvard University Press, 2004).

45. The citation for the 1965 report is this: "Rukovoditel' temi: incognito ispolniteli: khoteli byi ostat'sya neizvestnyimi, khotya byi dlya knachal'stva," *Otchyot laboratorii chitayushchikh avtomatov za 1964–1965 g.g.* (Kiev: Akademiya Nayk ukrainskoi ssr institute kibernetiki AN USSR, Laboratoriya chitayushchikh avtomatov, 1965).

46. Quote taken from unmarked document in personal archives of Vera Viktorevna Glushkova, Kiev, Ukraine.

47. Several scholarly works have drawn critical attention to the gendered performance in technical expertise, cyborg imagery, and counterculture, although none to my knowledge have done so in the Soviet context. For more, see Janet Abbate, *Recording Gender: Women's Changing Participation in Computing* (Cambridge: MIT Press, 2012); Donna Haraway, "A Cyborg Manifesto: Science, Technology, and Socialist-Feminism in the Late Twentieth Century," *Simians, Cyborgs and Women: The Reinvention of Nature* (New York: Routledge, 1991), 149–181; Nathan Engsmenger, *The Computer Boys Take Over: Computers, Programmers, and the Politics of Technical Expertise* (Cambridge: MIT Press, 2010); Turner, *From Counterculture to Cyberculture*, 76–77, 300–305.

48. In an unsent October 1961 letter, Nemchinov proposes an "institute of economic cybernetics." H. S. Khrushchyev, Doklad na XXII s'esde KPSS 18 Okctober 1961, RAN archives, CEMI, 1959, 1, 7, 124, 2.

49. Letter from Vasily Nemchinov to the Ministry of Finance: "V ministerstvo finansov SSSR" ["To the Ministry of Finance, USSR"], RAN archives, CEMI, 1959, 1, 7, 125, 11.

50. Ibid.

51. Kassel, *Soviet Cybernetics Research*, 94–96.

52. Nemchinov letter, "V ministerstvo finansov SSSR."

53. Ibid.

54. Yuri Gavrilets, interviewed by author, CEMI, Moscow, August 20, 2008.

55. Nemchinov letter, "V ministerstvo finansov SSSR."

56. Nikita S. Khrushchev, "Doklad na XXII S"esde KPSS" ["Concluding Speech"], *Twenty-second Congress*, October 18, 1961, accessed March 19, 2010, http://www.archive.org/details/DocumentsOfThe22ndCongressOfTheCpsuVolI.

57. Letter from V. S. Nemchinov to the Bureau of the Division of Economic, Philosophical, and Legal Sciences, Academy of Sciences, USSR, December 11, 1961, RAN Archives, 1959, 1, 7, 125, 11.

58. Report signed by V. S. Nemchinov, "Dokladnaya zapiska v Otdelenii, XXII s"ezd KPSS" ["Division Report Notes, Twenty-second Congress of the Communist Party"], November 17, 1961, RAN Archive, CEMI, 1959, 1, 6, 106.

59. Document signed by V. S. Nemchinov, "V Byuro otdeleniya ekonomicheskikh, filosophskikh I provavikh nauk AN CCCP" ["To the Office of the Division of Economic, Philosophical, and Legal Sciences, the Academy of Sciences, USSR"] September 17, 1960, RAN Archive, CEMI, 1960.

60. Nikolai Fedorenko, *Vestnik Akademii Nauk, SSSR* [*Herald of the Academy of Sciences, USSR*] 10 (1964): 3–14.

61. Kassel, *Soviet Cybernetics Research*, 98.

62. Interview with Yuri Gavrilets by the author, CEMI, Moscow, August 20, 2008. Data taken from the report titled "Doklad Akademika N. I. Fedoreko na yubileim zasedankii posvyashennoim 10-leniyu of Ts.E.M.I." ["Report by Academician N. I. Fedorenko on the Ten-Year Anniversary of CEMI"], CEMI archives, RAN, May 1973, 1959, 1, 403, 262.

63. Interview with Yuri Gavrilets by the author, CEMI, Moscow, August 20, 2008.

64. George Simmel's *Philosophie des Geldes* is a classic account of the form, not the value, of economic objects. As Simmel notes, "we may not describe exchangeability as a likeness of value that belongs objectively to things, but we must recognize likeness of value as simply a name for the exchangeability." George Simmel, *Philosophie des Geldes* [*The Philosophy of Money*] (Leipzig: Duncker & Humblot, 1900), 46.

65. Simmel writes, for example, that "we may not describe exchangeability as a likeness of value that belongs objectively to things, but we must recognize likeness of value as simply a name for the exchangeability." Ibid., 46.

66. Nikolai Fedorenko, *Vspominaya proshloe, vzglyadivaya v budushchee* [*Remembering the Past, Looking into the Future*] (Moscow: Nauka, 1999), 179.

67. Ibid., 179.

68. David Stark, *The Sense of Dissonance: Accounts of Worth in Economic Life* (Princeton: Princeton University Press, 2009), 19–31.

69. Fedorenko, *Vspominaya proshloe*, 209–214.

70. Informational document by Nikolai Fedorenko, "Istoricheskaya spravka o geyatel'nosti instituta s 1963 po 1966 g. by Director Akad. Fedorenko" ["Historical Information about the Activity of the Institute from 1963 to 1966 by Director Academician Fedorenko"], CEMI archives, RAN, 1959, 1, 101, index 170.

71. Kassel, *Soviet Cybernetics Research*, 87.

72. Homepage for Central Economic Mathematical Institute: "About CEMI" section, accessed April 15, 2015, http://www.cemi.rssi.ru/about/how/?section=about_link.

73. Eric P. Hoffmann and Robbin F. Laird. *Technocratic Socialism: The Soviet Union in the Advanced Industrial Era*. (Durham: Duke University Press, 1985), 116.

74. Ibid., 114.

75. Gerovitch, *From Newspeak to Cyberspeak*, 272.

76. Ibid.

77. Unsourced quote, accessed April 15, 2015, http://ogas.kiev.ua.

78. The original document proposing the EGSTVs in 1963 can be found here: Postanovlenie Komitet KPCC I Sovet Ministrov SSSR, "Ob uluchshenii rukovodstva vnedreniem vyichislitel'noi tekhniki i avtomatizirovannikh system upravleniya v narodnoe khozyaistvo," May 21, 1963, no. 564, Kremlin, Moscow. This document was published for the first time in Aleksei Viktorovich Kuteinikov, "Proekt Obschegosudarstvennoi avtomatizirovannoi sistemi upravleniya osvetskoi ekonomikoi (OGAS) I problem ego realizatii v 1960–1980-x gg," PhD dissertation, Moscow State University, Moscow, 2011. In addition, the 1967 proposal approving the regional Ukrainian GSTVs Project in 1967 can be found in Postanovleniya soveta ministrov Ukrainian SSR, "O vnedrenii avtomatizirovannikh system upravleniya s premeneniem vyichislitel'noi tekhniki," May 21, 1967, no. 338, and established by subsequent order of the Minister of Black Metallurgy, Ukrainian SSR, on March 11, 1968, no. 68.

79. Gerovitch, "InterNyet," 344. See also Iuliia Kapitonova and Aleksandr Letichevsky, *Paradigmy i idei akademika V. M. Glushkova* (Kiev: Naukova Dumka, 2003), 189.

80. Malinovskii, *Istoriia vychislitel'noi tekhniki*, 162.

81. Christopher Felix McDonald, "Building the Information Society: A History of Computing as a Mass Medium," Ph.D. diss., Princeton University, 2011.

82. Letter from the director of the CSA V. N. Starovskii to the director of the GK KNIR K. N. Rudnev, November 2, 1963, published in the appendix of Kuteinikov, "Proekt Obshchegosudarstvennoi."

83. V. P. Derkach, ed. *Akademik V. M. Glushkov—pioneer kiberniki* (Kiev: Yunior, 2003), 324.

84. Kathryn M. Bartol, "Soviet Computer Centres: Network or Tangle?," *Soviet Studies* 23 (4) (1972): 608–618.

85. Malinovsky, *Pioneers*, xxx–xxxii.

86. Ibid., 33.

87. Ibid., 165.

88. Malinovsky, *Store Eternally*, 61–62.

Chapter 5: The Undoing of the OGAS, 1970 to 1989

1. Gerovitch, "InterNyet," 343.

2. Glushkov, "Shto skazhet istoria?"

3. Malinovksy, *Vechno Khranit*, 61.

4. Gerovitch, *From Newspeak to Cyberspeak*, 280.

5. K. N. Rudnev, "Vyichislitel'naya tekhnika v narodnom khozyyaistve," *Izvestiya*, September 4, 1963, cited in Kuteinikov, "Pervie proekti," 136.

6. Malinovsky, *Istoriia vychislitel'noi tekhniki v litsakh* [*History of Computing Technology in Personalities*], reproduces a transcription of Glushkov's dictated memoirs, *Vopreki Avtoritetam* [*Despite the Authorities*], accessed April 15, 2015, http://lib.ru/MEMUARY/MALINOWSKIJ/5.htm, and in partial English translation in "Academician Glushkov's 'Life Work,'" accessed April 15, 2015, http://en.uacomputing.com/stories/ogas.

7. Glushkov, *Vopreki Avtoritetam*.

8. Ibid.

9. Ibid.

10. For more on the proposed structure of the OGAS, see Martin Cave, *Computers and Economic Planning: The Soviet Experience* (New York: Cambridge University Press, 1980), 13–15.

11. Malinovskii, *Istoriia vychislitel'noi tekhniki v litsakh*, 43–44.

12. Viktor M. Glushkov, "Dlya vsei strani," *Pravda*, December 13, 1981. See also Malinovsky, *Vechno Khranit*, 64, cf. 65. Other bibliographies suggest *strani* or "nation," not "state," is correct, although this remains unconfirmed.

13. Kuteinikov, "Pervie proekti," 97.

14. Gerovitch, "InterNyet," 345–346.

15. Kuteinikov, "Pervie proekti," 101.

16. Ibid., 119.

17. The ASUification (or *ASUchivaniye*) of the Soviet Union amounted to, the workers joked in Russian, the bitchification of the country because in Russian *ASUchivaniye* shares the same root as the swearword *suka*.

18. Malinovksy, *Istoriya vyichisletel'nikh tekhniki v litsakh*, 91, also 84–93.

19. Malinvosky, *Vechno Khranit*, 66–67.

20. Beissinger, *Scientific Management*, 249.

21. Glushkov, "Vopreki Avtoritetam."

22. Viktor Glushkov, "Ten Billion Accountants Needed," *RAND Report on Soviet Cybernetics* 2 (3) (1972): 72–73, accessed April 15, 2015, http://www.rand.org/content/dam/rand/pubs/reports/2007/R960.3.pdf.

23. Gerovitch, *From Newspeak to Cyberspeak*, 280.

24. Hughes, *Rescuing Prometheus*.

25. Eric P. Hoffmann and Robbin F. Laird, *Soviet Technocratic Socialism: The Soviet Union in the Advanced Industrial Era* (Durham: Duke University Press, 1985), 115.

26. Ibid., 115.

27. Ibid., 116.

28. Ibid., 116–117.

29. Ibid., 116.

30. Viktor Glushkov, "Zabetniye myislic dlya tekh, kto ostaetsya," January 10, 1982, *Akademik Glushkov—pioneer kibernetiki* (Kiev: n.p., 2003), accessed April 15, 2015, http://www.komproekt.ru/new/zavetnie_m.

31. Aleksandr Ivanovich Stavchikov, "Romantika pervyikh issledovannii i proektov i ikh protivorechnaya sud'ba" ["Romanticism of Early Research and Projects and Their Contradictory Fate"], in an unnamed, unpublished history of the Central Economic Mathematical Institute (TsEMI), chap. 2, Moscow, accessed 2008, 17. (See CEMI-RAS archive in bibliography.)

32. Stavchikov, "Romantika," 1–2.

33. Ibid., 16–17.

34. Ibid., 17.

35. In *The End of the Millennium*, Manuel Castells blames the incompatibility of a vertical statist hierarchy with horizontal information networks for the collapse of the Soviet Union, even while identifying ways that the Soviet Union did not behave as such a structure. Simultaneously, Gerovitch, in "InterNyet," claims that "Soviet cyberneticians envisioned an organic, self-regulating system, but paradoxically they

insisted on building it by decree from above." Although this claim is not wrong, it misses his earlier point that, aside from having no other option, the top-down system did not behave as a self-regulating hierarchy. The argument offered here looks to describe the same administrative challenges by using terms like *heterarchy*, which cuts a middle way through top-down and bottom-up, horizontal and vertical network structural discourse. Castells, *The End of the Millennium*, 26–37, 61–66; Gerovitch, "InterNyet," 347.

36. David Edmonds and John Eidinow's popular *Bobby Fisher Goes to War: How a Lone American Star Defeated the Soviet Chess Machine* (New York: Harper Perennial, 2005).

37. Zvi Y. Gitelman and Yaakov Ro'i, eds., *Revolution, Repression, and Revival: the Soviet Jewish Experience* (New York: Rowman & Littlefield, 2007), 119.

38. Frederic Bozo, Marie-Pierre Rey, N. Piers Ludlow, and Bernd Rother, eds., *Visions of the End of the Cold War in Europe, 1945–1990* (New York: Berghahn Books, 2012), 76–86.

39. Daniel Johnson, *White King and Red Queen: How the Cold War Was Fought on the Chessboard* (Boston: Houghton Mifflin, 2008), esp. chap. 6.

40. Boris Stillman, *Linguistic Geometry* (New York: Springer, 2000), xi.

41. Bruce Abramson, *Digital Phoenix: Why the Information Economy Collapsed and How It Will Rise Again* (Cambridge: MIT Press), 89–90.

42. Johnson, *White King and Red Queen*, chap. 6.

43. Nathan Engsmenger, "Is Chess the Drosophila of Artificial Intelligence?," *Social Studies of Science* 42 (1) (2011): 5–30. See also John McCarthy, "Chess as the Drosophila of AI," accessed April 15, 2015, http://jmc.stanford.edu/articles/drosophila/drosophila.pdf.

44. E. M. Landis and I. M. Yaglom, "About Aleksandr Semenovich Kronrod," *Uspekhi Matematicheskikh Nauk* 56 (5) (2001): 191–201, accessed April 15, 2015, http://www.mathnet.ru/links/1e483992e9f2c42fda4390d0116737a3/rm448.pdf.

45. Wiener, *God and Golem, Inc.*, 15–25.

46. Jad Abumrad and Robert Krulwich (hosts), "The Rules Can Set You Free," *RadioLab*, National Public Radio, April 9, 2013.

47. Walter Ong, *Orality and Literacy* (New York: Routledge, [1972] 2012), 82.

48. Philip von Hilger, *War Games: A History of War on Paper* (Cambridge: MIT Press, 2012).

49. Viktor Glushkov and V. Ya. Valakh, *Chto takoe OGAS?* (Moscow: Hauka, 1981), 1–160.

50. Malinovsky, *Vechno Khranit*, 57–58.

51. Letter from A. I. Kitov written on November 11, 1985, Politechnicheskii museum Russian Funderation, fond "Kitov Anatolii Ivanovich," f. 228, box BP 3450/1–2, reproduced in the appendix to Kuteinikov, "Project Obshchegosudarsvennoi."

52. Ibid.

53. Fedorenko, *Vspominaya Proshloe, Vzgladivaya v Budushchee*, 177.

54. Gerovitch, *From Newspeak to Cyberspeak*, 277.

55. Gerovitch, "Soviet InterNyet," 346; V. Golovachev, "A Hercules Is Born," *Soviet Cybernetics: Recent News Items* 5 (1967): 72.

56. Gerovitch, *From Newspeak to Cyberspeak*, 139.

57. Graham, *Lonely Ideas: Can Russia Compete?*

58. Gregory, *Restructuring the Soviet Economic Bureaucracy*, esp. introduction.

59. Kuteinikov, "Proyekt obshchegosudarstvennoi," 142.

60. Steven G. Medema, *The Hesitant Hand: Taming Self-Interest in the History of Economic Ideas* (Princeton: Princeton University Press, 2010), 6–10. See also Pierre Force, *Self-Interest before Adam Smith: A Genealogy of Economic Science* (New York: Cambridge University Press, 2003), which sees self-interest as a first principle behind what Hume calls the "selfish hypothesis" from the Epicureans through Jean Baptiste-Say.

61. Karl Eugen Wädekin, *The Private Sector in Soviet Agriculture* (Berkeley: University of California Press, 1973), xiv.

62. Dennis O'Hearn, "The Consumer Second Economy: Size and Effects," *Soviet Studies* 32 (2) (1980): 227, 232.

63. Gregory Grossman, "The Shadow Economy in the Socialist Sector of the USSR," in *The CMEA Five Year Plans (1981–1985) in a New Perspective: Planned and Non-Planned Economies*, ed. Economics and Information Directorate (Brussels: North Atlantic Treaty Organization, 1982), 108.

64. Gerald Mars and Yochanan Altman, *Private Enterprise in the USSR: The Case of Soviet Georgia* (Aldershot: Gower Press, 1987), chap. 6.

65. David Remnick, "Soviet Union's Shadow Economy: Bribery, Barter, and Black Market," *Seattle Times*, September 22, 1990, accessed April 15, 2015, http://community.seattletimes.nwsource.com/archive/?date=19900922&slug=1094485.

66. Seth Benedict Graham, "A Cultural Analysis of the Russo-Soviet 'Anekdot,'" Ph.D. diss., University of Pittsburgh, 2003; Remnick, "Soviet Union's Shadow Economy."

67. Two of the most classic cold war critics in the West include Friedrich Hayek, *The Road to Serfdom* (Chicago: University of Chicago, 1944), 4, 9–11, 28–29, 103–112, 143–145, and Milton Friedman, *Free to Choose: A Personal Statement* (New York: Harcourt Books, 1980). My comments here intend to both draw on and cut orthogonally across the conventional defense of free markets.

68. A sampling of popular literature on the informal economic activities in the late Soviet Union includes Yuri Brokhin, *Hustling on Gorky Street* (London: W. H. Allen, 1967); Konstantin Simis, *USSR: The Corrupt Society: The Secret World of Soviet Capitalism* (New York: Simon & Schuster, 1982); David Shipler, *Russia: Broken Idols, Solemn Dreams* (New York: Times Books, 1983); Lev Timofeev, *Soviet Peasants: Or the Peasants' Art of Starving* (New York: Telos Books, 1985). See also Gregory Grossman, ed., *Studies in the Second Economy of Communist Countries: A Bibliography* (Berkeley: University of California Press, 1988).

Conclusion

1. My thanks to Slava Gerovitch, who coined this pun in "InterNyet: Why the Soviet Union Did Not Build a Nationwide Computer Network." Others, such as Jack Balkin, have discovered it independently in conversation with the author as well. Barbara London named her 1998 video exhibit for the Museum of Modern Art, New York, "InterNyet: A Video Curator's Dispatches from Russia and Ukraine," accessed April 15, 2015, http://www.moma.org/interactives/projects/1998/internyet.

2. Graham, *Lonely Ideas*.

3. Jeff Weintraub, "The Theory and Politics of the Public/Private Distinction," in *Public and Private in Thought and Practice: Perspectives on a Grand Dichotomy*, ed. Jeff Weintraub and Krishan Kumar (Chicago: University of Chicago, 1997), 7, 8–16.

4. Arendt, *The Origins of Totalitarianism*.

5. Fedorenko, *Vspominaya proshloe, zaglyadyibayiu v budushchee*, 179.

6. Bonnie Honig, ed., *Feminist Interpretations of Hannah Arendt* (University Park: Pennsylvania State University Press, 1995); Mary Dietz, *Turning Operations: Feminism, Arendt, and Politics* (New York: Routledge, 2002).

7. Jonathan Coopersmith, "Failure and Technology," *Japan Journal for Science, Technology, and Society* 18 (2009): 93–118; Steven J. Jackson, "Rethinking Repair," in *Media Technologies: Essays on Communication, Materiality and Society*, ed. Tarleton Gillespie, Pablo Boczkowski, and Kirsten Foot, 221–240 (Cambridge: MIT Press, 2013).

8. Peter Galison, *How Experiments End* (Chicago: University of Chicago Press, 1987), 276.

9. Max Weber, *The Protestant Ethic and the Spirit of Capitalism* (Boston: Unwin Hyman, [1905] 1930); see also R. H. Howe, "Max Weber's Elective Affinities," *American Journal of Sociology* 84 (1978): 366–385; Ludwig Fleck, *Genesis and Development of a Scientific Fact*, trans. Fred Bradley and Thaddeus J. Trenn, ed. Thaddeus J. Trenn and Robert K. Merton (Chicago: Chicago University Press, [1935] 1979).

10. Two researchers have characterized the stereotypical difference between Russian and Chinese informal influence as trending toward a logical-analytic mindset and a holistic-dialectical one. Snejina Michailova and Verner Worm, "Personal Networking in Russia and China: *Blat* and *Guanxi*," *European Management Journal* 21 (4) (2003): 509–519.

11. For more on technological utopianism in global contexts, see Howard P. Segal *Technology and Utopia* (Washington, DC: American Historical Association, 2006); for more on the Swedish Pirate Party, see Patrick Burkart, *Pirate Politics: The New Information Policy Contests* (Cambridge: MIT Press, 2014).

12. Michael Gordin, Hellen Tilley, and Gyan Prakash, "Introduction," in *Utopia/ Dystopia: Conditions of Historical Possibility*, ed. Michael D. Gordin, Helen Tilley, and Gyan Prakash (Princeton: Princeton University Press, 2010), 2, see also 1–6.

13. Cat video scholarship exists. See Jody Berland, "Cat and Mouse: Iconographics of Nature and Desire," *Cultural Studies* 22 (3–4) (2008): 431–454.

14. Jürgen Habermas, *The Structural Transformations of the Public Sphere: An Inquiry into a Category of Bourgeois Society* (Cambridge: MIT Press, 1989).

15. Nicholas John and Benjamin Peters, "Is the End Always Near? An Analysis and Comment on the End of Privacy, 1990–2012," unpublished manuscript; Daniel J. Solove, "A Taxonomy of Privacy," *University of Pennsylvania Law Review* 154 (3) (2006): 477–560.

16. George L. Priest, "The Ambiguous Moral Foundations of the Underground Economy," *Faculty Scholarship Series*, Paper 626 (1995), accessed April 15, 2015, http:// digitalcommons.law.yale.edu/cgi/viewcontent.cgi?article=1625&context=fss _papers.

17. Eric Schatzberg, "Technik Comes to America: Changing Meanings of Technology before 1930," *Technology and Culture* 47 (2006): 486–511; see also Martin Heidegger, *The Question concerning Technology and Other Essays*, trans. William Lovitt (New York: Harper & Row, [1954] 1977), 287–317.

18. Raymond Williams, "Base and Superstructure in Marxist Cultural Theory," *New Left Review* 1 (82) (1973): 1–13; Raymond Williams, "Culture Is Ordinary," in *Resources of Hope: Culture, Democracy, Socialism* (London: Verso), 3–14; Raymond Williams, *Television: Technology and Cultural Form* (London: Wesleyan University Press, [1974] 1992): 1–25.

Bibliography

Primary Sources

The following primary sources were consulted in writing this book and are organized below by archive.

Archives

Central Economic-Mathematical Institute, Russian Academy of Sciences (CEMI-RAS) Nemchinov letter to N. S. Khrushchyev, Doklad na XXII s'esde KPSS, October 18, 1961. RAS archives, CEMI, 1959, 1, 7, 124, 2.

Vasily Nemchinov to the Ministry of Finances, USSR: "V ministerstvo finansov SSSR" ["To the Ministry of Finances, USSR"]. RAS archives, CEMI, 1959, 1, 7, 125, 11.

Letter signed by V. S. Nemchinov dated December 11, 1961, to the "Bureau of the Division of Economic, Philosophical, and Legal Sciences, Academy of Sciences, USSR." RAS Archives, 1959, 1, 7, 125, 11.

Report signed by V. S. Nemchinov, "Dokladnaya zapiska v Otdelenii, XXII s'ezd KPSS" ["Division Report Notes, Twenty-second Congress of the Communist Party"], November 17, 1961. RAS Archive, CEMI, 1959, 1, 6, 106.

Document signed by V. S. Nemchinov, "V Byuro otdeleniya ekonomicheskikh, filosophskikh i provavikh nauk AN CCCP," [To the Office of the Division of Economic, Philosophical, and Legal Sciences, Academy of Sciences, USSR] September 17, 1960. RAS Archive, CEMI, 1960 Document.

"Doklad Akademika N. I. Fedoreko na yubileim zasedankii posvyashennoim 10-leniyu of Ts.E.M.I." ["Report by Academician N. I. Fedorenko on the Ten-Year Anniversary of CEMI"], May 1973. CEMI archives, RAS, 1959, 1, 403, 262.

Informational document by Nikolai Fedorenko, "Istoricheskaya spravka o geyatel'nosti instituta s 1963 po 1966 g. by Director Akad. Fedorenko" ["Historical

Information about the Activity of the Institute from 1963 to 1966 by Director Academician Fedorenko"]. CEMI archives, RAS, 1959, 1, 101, index 170,

Aleksandr Ivanovich Stavchikov, academic secretary of CEMI, "Romantika pervyikh issledovanii i proektov i ikh protivorechnaya sud'ba" ["Romanticism of Early Research and Projects and Their Contradictory Fate"], unnamed, unpublished history of Central Economic Mathematical Institute (CEMI), Moscow, read in person and returned May 2008, chap. 2, 17.

Central Intelligence Agency (CIA)   *These materials were obtained by Freedom of Information Act requests.*

Ford, John J. "Artificial Limbs." Scientific and Technical Intelligence Report No. 464691. May 10, 1967.

Ford, John J. "The Cybernetic Approach to Education in the USSR." Scientific Intelligence Memorandum No. 464693. May 25, 1964.

Ford, John J. "Major Developments in the SovBloc Cybernetics Programs in 1965." Scientific and Technical Intelligence Report No. 464694. October 3, 1966, 1–33.

Ford, John J. "The Meaning of Cybernetics in the USSR." Intelligence Memorandum No. 9757/64. February 26, 1964, 1–10, esp. 1.

Ford, John J. "The Soviet Applications of Cybernetics in Medicine: 1. Medical Diagnosis." Scientific and Technical Intelligence Report No. 464692. September 15, 1966.

Kuteinikov, Aleksei Viktorovich. "Proekt avtomatizirovannoi sistemi upravleniya sovetskoi ekonomikoi (OGAS) i problem ego realizatsii v 1960–1980-x gg." ["The project of all-state automated system of management of Soviet economics (OGAS) and the problem of its realization in 1960-1980"] Ph.D. diss., Moscow State University, 2011.

Dissertations   *The following primary documents are reproduced in the appendix to Kuteinikov's dissertation.*

Kitov, Anatoly. "Pis'mo zamestitelya nachal'nika VTs Minoboroni SSSR A. U. Kitova v TsK KPSS N. S. Khrushchyovu ot 7 Anvarya 1959 goda." [Letter from the Vice Chief of Computing Center for the Ministry of Defense USSR A. U. Kitov to the General Secretary of the Communist Party N.S. Khrushchev on January 7, 1959.] Politekhnicheskii Museum of the Russian Federation, fond "Kitov Anatolii Ivanovich," f. 228, edinitsa khraneniya KP27189/20.

Kitov, Anatoly. November 11, 1985, Politechnicheskii Museum of the Russian Federation, fond "Kitov Anatolii Ivanovich," f. 228, box BP 3450/1–2.

Postanovlenie Tsentralnogo Komiteta KPCC i Soveta Ministrov SSSR. "Ob uluchshenii rukovodstva vnedreniem bychislitel'noi tekhniki i avtomatizirovannikh system upravleniya v narodnoe khozyaistvo" ["On the improvement of the leadership over the introduction of computing technology and automated systems of management in the national economy"], May 21, 1963, no. 564, Gosudarstvenni archive (GARF), f. 5446, o. 106, d. 1324, l. 160–172.

Kovalev, N. I. "Doklad o rabote i perspektivakh razvitiya VTs pri Gosekonomsovete, 23 iyulya 1962 g." ["Report about the work and perspectives of the development of computing networks in the State Economics Committee, 23 July 1962"] Rossiiskii gosudarstvennyi arkhiv ekonomiki (RGAE). f. 9480, o. 7, d. 466, l. 77–97.

Postanovleniya soveta ministrov Ukrainian SSR, May 27, 1967, no. 338. "O vnedrenii avtomatizirovannikh system upravleniya s premeneniem vyichislitel'noi tekhniki" ["On the introduction of automated systems of management with the application of computing technology"]. Established by subsequent order of the Minister of Black Metallurgy, Ukrainian SSR, on March 11, 1968, no. 68.

Starovskii, V. N. Letter from the director of the CSA USSR, V. N. Starovskii, to the director of the GK KNIR, K. N. Rudnev, November 2, 1963.

"Voprosi strukturi, organizatsii i sozdaniya edinoi gosudarstvennoj seti vyichislitel'nikh tsentrov EGSVTs" ["Questions of the structure, organization, and creation of a unified state network of computing centers, EGSVTs"]. *Rossiiskii gosudarstvennyi arkhiv ekonomiki* (RGAE). f. 9480, o. 7, d. 1227, l. 82–102.

National Academy of Sciences, Ukraine (NAS) Glushkov, Viktor M. National Academy of Sciences of Ukraine. Archive and Special Collections, Viktor M. Glushkov's personal files. box 18, folder 1, documents 12, 119, 122, and 123 inclusive.

Peters, Benjamin *These documents are held only in the author's personal archive.*

"Podorozh v Krainu Kibertonii" ["Travel to the Country of Cybertonia"]. *Kievskii Komsomoltsyi*, August 1, 1963, 2–3.

"Rukovoditel' temi: incognito ispolniteli: khoteli byi ostat'sya neizvestnyimi, khotya byi dlya nachal'stva" ["Incognito Masters: (on) wanting to remain invisible at least to supervisors"]. *Otchyot laboratorii chitayushchikh avtomatov za 1964–1965 g.g.* Kiev: Akademiya Nayk ukrainskoi ssr institute kibernetiki AN USSR, Laboratoriya chitayushchikh avtomatov, 1965.

"Vechirniy Kiev." ("Evening Kiev") 305 (5624), December 31, 1962, 3.

"Vechirniy Kiev." ("Evening Kiev") 309 (6588), December 31, 1965, 2–3.

"Vecherniy Kyber" ("Evening Cyber: Newspaper of the Council of Robots"). 1 (1) (1966).

Voloshin, A. "Kibertonia-65." ("Cybertonia-65") *Vechirnii Kiev*, February 16, 1965, 2.

Other Archival Sources Consulted

Central Economic Mathematical Institute, Moscow, Russia

Glushkov, Viktor, uncollected archives in the closet of Glushkov's former main office, Institute of Cybernetics, Room 804, Kiev, Ukraine.

Glushkova, Vera Viktorevna, Kiev, Ukraine, personal archives

Institute of Cybernetics, Kiev, Ukraine

National Academy of Sciences of Ukraine, Kiev, Ukraine

National Library of Ukraine, Kiev, Ukraine

Russian Academy of Sciences (RAS), Moscow, Russia

Interviews with OGAS Project Consultants

The following interviews were conducted by the author in Russian in Moscow, Russia, and Kiev, Ukraine. Although not always directly involved in the OGAS Project, these researchers were or are associated with relevant institutions and social networks.

Gavrilets, Yuri N. August 20, 2008, Moscow, Russia

Goshka, Vladimir. May 13, 2012, Kiev, Ukraine

Glushkova, Vera Viktorevna. May 14 and 30, 2012 (and after), Kiev, Ukraine

Grinchuk, Volodimir Mihailovich. May 18, 2012, Kiev, Ukraine

Karchenets, Dmitri Vasilivich. May 16, 2012, Kiev, Ukraine

Klimenko, Vitaliy Petrovych. May 18, 2012, Kiev, Ukraine

Kuteinikov, Aleksei Viktorovich. May 7–15, 2012 (and after), Kiev, Ukraine

Makaraov, Valerii Leonidovich. August 19, 2008, Moscow, Russia

Malinovskii, Boris Nikolaevich. May 7 and 18, 2012, Kiev, Ukraine

Morozov, A. A. May 13, 2012, Kiev, Ukraine

Nekrasova, Leonina Nikolaevna. February 23, 2012, Kiev, Ukraine

Palagin, A. V. May 9, 2012, Kiev, Ukraine

Roenko, Nikolai V. May 17, 2012, Kiev, Ukraine

Sergienko, I. V. May 6, 2012, Kiev, Ukraine

Stavchikov, Aleksandr Ivanovich. August 24, 2008, Moscow, Russia

Suschenko, Aleksandr. May 10 and 13, 2012, Kiev, Ukraine

Zhabin, Sergei. May 7–May 13, 2012, Kiev, Ukraine

Bibliography

Abbate, Janet. *Inventing the Internet*. Cambridge: MIT Press, 1999.

Abbate, Janet. *Recording Gender: Women's Changing Participation in Computing*. Cambridge: MIT Press, 2012.

Abramson, Bruce. *Digital Phoenix: Why the Information Economy Collapsed and How It Will Rise Again*. Cambridge: MIT Press, 2005.

Adamsky, Dima. *The Culture of Military Innovation: The Impact of Cultural Factors on the Revolution in Military Affairs in Russia, the US, and Israel*. Stanford: Stanford University Press, 2010.

Arendt, Hannah. *The Origins of Totalitarianism*. New York: Schocken Books, 1951.

Armstrong, George M., Jr. *The Soviet Law of Property*. The Hague: Martinus Nijhoff, 1983.

Ariely, Dan. *Predictably Irrational: The Hidden Forces That Shape Our Decisions* (New York: HarperCollins, 2008)

Armstrong, John A. "Sources of Administrative Behavior: Some Soviet and Western European Comparisons." *American Political Science Review* 59 (3) (1965): 643–655.

Aslund, Anders. *How Capitalism Was Built: The Transformation of Central and Eastern Europe*. New York: Cambridge University Press, 2007.

Aspray, William. *John von Neumann and the Origins of Modern Computing*. Cambridge: MIT Press, 1990.

Aspray, William. "The Scientific Conceptualization of Information." *Annals of the History of Computing* 7 (2) (1985): 117–140.

Aspray, William, and Paul E. Ceruzzi. *The Internet and American Business*. Cambridge: MIT Press, 2008.

Baran, Paul. "Introduction to Distributed Communication Networks." *On Distributed Communications*. RAND Corporation Memorandum RM-3420-PR, August 1964.

Bartol, Kathryn M. "Soviet Computer Centres: Network or Tangle?" *Soviet Studies* 23 (4) (1972): 608–618.

Becker, Abraham S. "Input-Output and Soviet Planning: A Survey of Recent Developments." Memorandum prepared for the United States Air Force, RAND Memorandum RM 3523-PR, March 1963. Accessed July 18, 2013, http://www.dtic.mil/cgi-bin/GetTRDoc?AD=AD0401490.

Beer, Stafford. *Brain of the Firm*. London: Allen Lane, Penguin Press, 1972.

Beissinger, Mark. *Scientific Management, Socialist Discipline, and Soviet Power*. Cambridge: Harvard University Press, 1988.

Benkler, Yochai. *The Wealth of Networks: How Social Production Transforms Markets and Freedom*: New Haven: Yale University Press, 2006.

Berg, Aksel' I., A. I. Kitov, and A. A. Lypunov. "O vozmozhnostyakh avtomatizatsii upravleniya narodniym kozyaistvom" ["On the possibilities of automating the management of the national economy"]. Moscow, November 1959. Accessed July 25, 2013, http://www.kitov-anatoly.ru/naucnye-trudy/izbrannye-naucnye-trudy-anatolia-ivanovica-v-pdf/o-vozmoznostah-avtomatizacii-upravlenia-narodnym-hozajstvom.

Bergier, Jacque. "Un plan général d'automatisation des industries," *Les Lettres françaises* (April 15, 1948): 7–8.

Berland, Jody. "Cat and Mouse: Iconographics of Nature and Desire." *Cultural Studies* 22 (3–4) (2008): 431–454.

Berlin, Isaiah. *Russian Thinkers*. New York: Penguin Group, 1978.

Biggart, J., P. Dudley, and F. King, eds. *Alexander Bogdanov and the Origins of Systems Thinking in Russia*. Brookfield, VT: Ashgate, 1998.

Billington, James H. *The Icon and the Axe: An Interpretive History of Russian Culture*. New York: Vintage, 1966.

Birman, A. "Neotvratimost" ["Inevitability"]. *Zvezda* 5 (1978).

Blau, Peter. *Bureaucracy in Modern Society*. New York: Random House, 1956.

Boczkowski, P., and Lievrouw, L. "Bridging STS and Communication Studies: Scholarship on Media and Information Technologies." In *The Handbook of Science and Technology Studies*, 3rd ed. Edited by E. Hackett, O. Amsterdamska, M. Lynch, and J. Wajcman, 949–977. Cambridge: MIT Press, 2007.

Borovitski, I. Editorial, *Pravda*, October 3, 1962.

Bogdanov, Aleksandr. *Tektologia: Vsyeobshcheiye Organizatsionnaya Nauka* [*Tectology: A Universal Organizational Science*]. Moscow: Akademia Nauk, 1913–1922.

Bowker, Geoffrey C. "How to Be Universal: Some Cybernetic Strategies, 1943–1970." *Social Studies of Science* 23 (1993): 107–127.

Bowker, Geoffrey C. "The Empty Archive: Cybernetics and the 1960s," in *Memory Practices in the Sciences* (Cambridge, MA: MIT Press, 2006).

Bowker, Geoffrey C., and Leigh Starr. *Sorting Things Out: Classification and Its Consequences*. Cambridge: MIT Press, 1999.

Bozo, Frédéric, Marie-Pierre Rey, N. Piers Ludlow, and Bernd Rother. *Visions of the End of the Cold War in Europe, 1945–1990*. New York: Berghahn Books, 2012.

Brand, Stewart. "Founding Father." *Wired* 9 (3) (1991). Accessed April 15, 2015, http://archive.wired.com/wired/archive/9.03/baran_pr.html.

Brillouin, Léon. "Les grandes machines mathématiques américaines," *Atomes* 2 (21) (1947): 400–404.

Brillouin, Léon. "Les machines américaines," *Annales des Télécommunications* 2 (1947): 331–346.

Broglie, Louis de. *La cybernétique: théorie du signal et de l'information.* Paris: Edition de la Revue d'Optique Théorique et Instrumentale, 1951.

Brokhin, Yuri. *Hustling on Gorky Street.* London: Allen, 1967.

Brown, Peter. *The Body and Society: Men, Women, and Sexual Renunciation in Early Christianity.* New York: Columbia University Press, 1988.

Bruk, Isaak C. "Elektronnie vyichislitel'nie mashinyi—na sluzhbu narodnomu khozyaistvu" ["Electronic computing machines—in the service of the national economy"], *Kommunist* 7 (1957).

Bryld, Mette, and Nina Lykke. *Cosmodolphins: Feminist Cultural Studies of Technology, Animals and the Sacred.* New York: Zed Books, 2000.

Brzezinski, Zbigniew K. *The Soviet Block: Unity and Conflict.* Rev. ed. New York: Praeger, 1960.

Burkart, Patrick. *Pirate Politics: The New Information Policy Contests.* Cambridge: MIT Press, 2014.

Burton, Finn. *Spam: A Shadow History of the Internet.* Cambridge: MIT Press, 2013.

Business Week, "The Office of the Future," *Business Week* 30 (2387) (1975): 48–70.

Campbell-Kelly, Martin, and William Aspray. *Computer: A History of the Information Machine.* Boulder, CO: Westview, 2004.

Carey, James W. "A Cultural Approach to Communication." In *Communication as Culture: Media and Society.* Edited by James Carey, 13–36. New York: Unwin Hyman, 1989.

Carey, James W., with John J. Quirk. "The Mythos of the Electronic Revolution." In *Communication and Culture: Essays on Media and Society.* Edited by James W. Carey, 113–141. New York: Unwin Hyman, 1989.

Carter, Ashton B., and David N. Schwartz. *Ballistic Missile Defense.* Washington, DC: Brookings Institution Press, 1984.

Castells, Manuel. *End of the Millennium: The Information Age—Economy, Society, and Culture.* Oxford: Blackwell, 1998.

Cave, Martin. *Computers and Economic Planning: The Soviet Experience.* New York: Cambridge University Press, 1980.

Central Economic-Mathematical Institute (CEMI). Homepage. "About CEMI" section. Accessed April 15, 2015, http://www.cemi.rssi.ru/about/how/?section=about_link.

Ceruzzi, Paul E. *Computing: A Concise History*. Cambridge: MIT Press, 2012.

Ceruzzi, Paul E. *A History of Modern Computing*. Cambridge: MIT Press, 1998.

Chapek, Karel. *Rossum's Universal Robots*. Translated by David Willie. Fairford: Echo Library, 2010.

Chesnenko, N. "'Obshchii iazyk' elektronnykh mashin: Problemy kodirovaniia dannykh" ["The general language of electronic machines: the problems of coding data"], *Ekonomiceskaia gazeta* (47) (1973): 10.

Coase, Ronald. "The New Institutional Economics." *American Economic Review* 88 (2) (1998): 72–74.

Cohen, Stephen F. *Bukharin and the Bolshevik Revolution: A Political Biography, 1888–1938*. New York: Oxford University Press, 1971.

Conquest, Robert. *The Harvest of Sorrow*. New York: Oxford University Press, 1986.

Conway, Flo, and Jim Siegelman. *Dark Hero of the Information Age: In Search of Norbert Wiener, the Father of Cybernetics*. New York: Basic Books, 2005.

Cooley, Charles Horton. *Sociological Theory and Social Research: Selected Papers of Charles Horton Cooley*. New York: Holt, 1930.

Coopersmith, Jonathan. "Failure and Technology." *Japan Journal for Science, Technology and Society* 18 (2009): 93–118.

Crowther-Heyck, Hunter. *Herbert A. Simon: The Bounds of Reason in Modern America*. Baltimore: Johns Hopkins University Press, 2005.

Davenport, Lisa E. *Jazz Diplomacy: Promoting America in the Cold War Era*. Jackson: University Press of Mississippi, 2009.

DeJong-Lambert, William, and Nikolai Krementsov. "On Labels and Issues: The Lysenko Controversy and the Cold War." *Journal of the History of Biology* 45 (3) (2012): 373–388.

Derkach, V. P., ed. *Akademik V. M. Glushkov—pioneer kiberniki [Academician V. M. Glushkov: Pioneer of Cybernetics]*. Kiev: Yunior, 2003.

Dewey, John. *The Essential Dewey*. Vol. 2. Bloomington: Indiana University Press, 1999.

Dietz, Mary. *Turning Operations: Feminism, Arendt, and Politics*. New York: Routledge, 2002.

Divine, Robert A. *The Sputnik Challenge: Eisenhower's Response to the Soviet Satellite*. New York: Oxford University Press, 1993.

Dolgov, V. A. *Kitov Anatolii Ivanovich—pioner kibernetiki, informatiki I avtomatizirovannikh system upravleniya: Nauchno-biograficheskii ocherk. [Kitov, Anatolii*

Ivanovich—*Pioneer of cybernetics, informatics, and automated systems of management: a scientific-biographical essay.]* Moscow: KOS-INF, 2010.

Dolot, Miron. *Execution by Hunger: The Hidden Holocaust.* New York: Norton, 2011.

Draganescu, Mihai. *Odobleja: Between Ampère and Wiener.* Bucharest: Academia Republicii Socialiste Romania, 1981.

Dubarle, Dominique. "Idées scientifiques actuelles et domination des faits humains," *Esprit* 9 (18) (1950): 296–317.

Dubarle, Dominique. "Une nouvelle science: la cybernétique—vers la machine à gouverner?," *Le Monde*, December 28, 1948, in P. Breton, *A l'image de l'homme* (Paris: Seuil, 1995), 137–138.

Dupuy, Jean-Pierre. *The Mechanization of the Mind: The Origins of Cognitive Science.* Translated by M. B. DeBevoise. Cambridge: MIT Press, 2009.

Dyker, David. *Restructuring the Soviet Economy.* New York: Routledge, 1991.

Dyson, George. *Turing's Cathedral: The Origins of the Digital Universe.* Pantheon Books: New York, 2012.

Eames, Charles, and Ray Eames. *A Computer Perspective: Background to the Computer Age.* Cambridge: Harvard University Press, 1973.

Edmonds, David, and John Eidinow. *Bobby Fisher Goes to War: How a Lone American Star Defeated the Soviet Chess Machine.* New York: Harper Perennial, 2005.

Edwards, Paul N. *The Closed World: Computers and the Politics of Discourse in Cold War America.* Cambridge: MIT Press, 1996.

Ellman, Michael. *Planning Problems in the USSR: The Contributions of Mathematical Economics to their Solution, 1960–1971.* New York: Cambridge University Press, 1973.

Ellman, Michael. *Socialist Planning.* New York: Cambridge, 1978.

Engels, Friedrich. "Letter to Karl Kautsky in Vienna." In Karl Marx and Frederick Engels, *Selected Correspondence.* Moscow: Progress, 1975. Accessed July 25, 2013, http://www.marxists.org/archive/marx/works/1881/letters/81_02_01.htm.

Engerman, David C. *Modernization from the Other Shore: American Intellectuals and the Romance of Russian Development.* Cambridge: Harvard University Press, 2003.

Engsmenger, Nathan. *The Computer Boys Take Over: Computers, Programmers, and the Politics of Technical Expertise.* Cambridge: MIT Press, 2010.

Engsmenger, Nathan. "Is Chess the Drosophila of Artificial Intelligence? A Social History of an Algorithm." *Social Studies of Science* 42 (1) (2011): 5–30.

Erickson, Paul, Judy L. Klein, Lorraine Dastone, Rebecca Lemov, Thomas Sturm, and Michael D. Gordin. *How Reason Almost Lost Its Mind: The Strange Career of Cold War Rationality.* Chicago: University of Chicago Press, 2013.

Ericson, Richard E. "Command Economy," *The New Palgrave Dictionary of Economics*, 2nd ed. Edited by Steven N. Durlauf and Lawrence E. Blume. New York: Palgrave, 2008.

Federenko, Nikolai. *Vestnik Akademii Nauk, SSSR* 10 (1964): 3–14.

Fedorenko, Nikolai. *Vspominaya proshloe, vzglyadivaya v budushchee* [*Remembering the Past, Looking into the Future*]. Moscow: Nauka, 1999.

Ferguson, Niall. *The Ascent of Money: A Financial History of the World*. New York: Penguin Group, 2008.

Figes, Orlando. *Natasha's Dance: A Cultural History of Russia*. New York: Picador, 2003.

Figes, Orlando. *Revolutionary Russia, 1891–1991: A History*. New York: Metropolitan Books, 2014.

Firth, Noel E., and James H. Noren, *Soviet Defense Spending: A History of CIA Estimates, 1950–1990*. College Station: Texas A&M University Press.

Fleck, Ludwig. *Genesis and Development of a Scientific Fact*. Translated by Fred Bradley and Thaddeus J. Trenn. Edited by Thaddeus J. Trenn and Robert K. Merton. Chicago: University of Chicago Press, 1979.

Force, Pierre. *Self-Interest before Adam Smith: A Genealogy of Economic Science*. New York: Cambridge University Press, 2003.

Friedman, Milton. *Free to Choose: A Personal Statement*. New York: Harcourt Books, 1980.

Gaddis, John Lewis. *The Cold War: A New History*. New York: Penguin Press, 2005.

Galison, Peter. *How Experiments End*. Chicago: University of Chicago Press, 1987.

Galison, Peter. *Image and Logic: A Material Culture of Microphysics*. Chicago: University of Chicago Press, 1997.

Galison, Peter. "The Ontology of the Enemy: Norbert Wiener and the Cybernetic Vision." *Critical Inquiry* 21 (1) (1994): 228–266.

Galloway, Alex. *Protocol: How Control Exists after Decentralization*. Cambridge: MIT Press, 2004.

Gardner, Roy. "L. V. Kantorovich: The Price Implications of Optimal Planning." In *Socialism and the Market: Mechanism Design Theory and the Allocation of Resources*. Edited by Peter J. Boettke. New York: Routledge, 2000.

Geoghegan, Bernard. *The Cybernetic Apparatus: Media, Liberalism, and the Reform of the Human Sciences*. Ph.D. diss., Northwestern University, 2012.

Geoghegan, Bernard. "From Information Theory to French Theory: Jakobson, Lévi-Strauss, and the Cybernetic Apparatus." *Critical Inquiry* 38 (2011): 96–126.

Geoghegan, Bernard. "The Historiographic Conceptualization of Information: A Critical Survey." *IEEE Annals of the History of Computing* 30 (2008): 66–81.

Geoghegan, Bernard, and Benjamin Peters, "Cybernetics." In *The John Hopkins Guide to Digital Media.* Edited by by Marie-Laure Ryan, Lori Emerson, and Benjamin J. Robertson. Baltimore: John Hopkins University Press, 2014.

Gerovitch, Slava. "The Cybernetics Scare and the Origins of the Internet." *Baltic Worlds* 2 (1) (2009): 32–38.

Gerovitch, Slava. *From Newspeak to Cyberspeak: A History of Soviet Cybernetics.* Cambridge: MIT Press, 2002.

Gerovitch, Slava. "Human-Machine Issues in the Soviet Space Program." *Critical Issues in the History of Spaceflight.* Edited by Steven J. Dick and Roger D. Launius, 107–140. Washington, DC: NASA History Division, 2006.

Gerovitch, Slava. "InterNyet: Why the Soviet Union Did Not Build a Nationwide Computer Network." *History and Technology* 24 (4) (2008): 335–350.

Gerovitch, Slava. "Speaking Cybernetically: The Soviet Remaking of an American Science." Ph.D. diss., Program in Science, Technology and Society, Massachusetts Institute of Technology, 1999.

Gitelman, Zvi Y., and Yaakov Ro'i, eds. *Revolution, Repression, and Revival: The Soviet Jewish Experience.* New York: Rowman & Littlefield, 2007.

Glushkova, Vera V. "Dorogoi chitatel', dobro pozhalovat' v 'kibertoniyu'!" ["Dear Reader, Welcome to Cybertonia!"] *Cybertonia* 1 (1) (2012): 2. Accessed April 15, 2015, http://miratechgroup.com/sites/default/files/documents/press_about_us/kibertonia_n01-2012.pdf.

Glushkova, Vera V., and Sergei A. Zhabin. "Virtualnaya strana Kibertonia v institute kibernetiki (60–70 gg, XX vek) [The virtual country Cybertonia in the institute of cybernetics (1960s-1970s)." *Ukrainia i svit: gumanitarno-tekhnicheska elita ta sotsialnyi progress: tezi dopov* 3, 81–83. Kharkiv: NTU Kharkiv, 2012.

Glushkov, Viktor. "Dlya vsei strain" ["For the whole country"]. *Pravda,* December 13, 1981.

Glushkov, Viktor. "Kibernetika, progress, budushchee" ["Cybernetics, progress, the future"]. *Literaturnaya Gazeta,* September 25, 1962.

Glushkov, Viktor. *Kibernetika, vychislitel'naia tekhnika, informatika. Izbrannye trudy [Cybernetics, computing technology, informatics. Selected works],* 2. Kiev: Naukova Dumka, 1990.

Glushkov, Viktor. OGAS. Unsourced quote. Accessed April 15, 2015, http://ogas.kiev.ua.

Glushkov, Viktor. *Osnovi bezbumazhnoi informatiki [The foundations of paperless informatics]*. Moscow: Nauka, 1982.

Glushkov, Viktor. "Shto skazhet istoria" ["What history will tell"], 3. Accessed April 15, 2015, http://ogas.kiev.ua/history/chto-skazhet-ystoryya.

Glushkov, Viktor. "Ten Billion Accountants Needed." *RAND Report on Soviet Cybernetics* 2 (3) (1972): 72–73. Accessed April 15, 2015, http://www.rand.org/content/dam/rand/pubs/reports/2007/R960.3.pdf.

Glushkov, Viktor. "Vopreki avtoritetam" ["Despite the authorities"]. Accessed April 15, 2015, http://lib.ru/MEMUARY/MALINOWSKIJ/5.htm.

Glushkov, Viktor. "Zavetniye myislic dlya tekh, kto ostaetsya" ["Treasured thoughts for those who remains"] (January 10, 1982). *Akademik Glushkov—pioneer kibernetiki* (Kiev: n.p, 2003). Accessed April 15, 2015, http://www.komproekt.ru/new/zavetnie_m.

Glushkov, Viktor, and V. Ya. Valakh. *Chto takoe OGAS? [What is OGAS?]* Moscow: Nauka, 1981.

Golovachev, V. "A Hercules Is Born." *Soviet Cybernetics: Recent News Items* 5 (1967): 70–78.

Gordin, Michael, Hellen Tilley, and Gyan Prakash. "Introduction." In *Utopia/Dystopia: Conditions of Historical Possibility*. Edited by Michael D. Gordin, Helen Tilley, and Gyan Prakash. Princeton: Princeton University Press, 2010.

Gould, Stephen Jay. *Life's Grandeur*. London: Vintage, 1997.

Graeber, David. *Debt: The First Five Thousand Years*. New York: Melville House, 2011.

Graham, Loren R. *The Ghost of the Executed Engineer: Technology and the Fall of the Soviet Union*. Cambridge: Harvard University Press, 1993.

Graham, Loren R. "How Robust Is Science under Stress?" *What Have We Learned about Science and Technology from the Russian Experience?*, 52–73. Stanford: Stanford University Press, 1998.

Graham, Loren R. *Lonely Ideas: Can Russia Compete?* Cambridge: MIT Press, 2013.

Graham, Loren R. *Science in Russia and the Soviet Union: A Short History*. New York: Cambridge University Press, 1993.

Graham, Loren R. *Science, Philosophy, and Human Behavior in the Soviet Union*. New York: Columbia University, 1987.

Graham, Loren, and Jean-Michael Kantor, *Naming Infinity: A True Story of Religious Mysticism and Mathematical Creativity*. Cambridge: Belknap Press of Harvard University Press, 2009.

Graham, Seth Benedict. "A Cultural Analysis of the Russo-Soviet 'Anekdot.'" Ph.D. diss., University of Pittsburgh, 2003.

Granick, David. *Management of the Industrial Firm in the USSR: A Study in Soviet Economic Planning*. New York: Columbia University Press, 1955.

Granovetter, Mark. "The Strength of Weak Ties." *American Journal of Sociology* 78 (6) (1973): 1360–1380.

Greenhalgh, Susan. "Missile Science, Population Science: The Origins of China's One-Child Policy." *China Quarterly* 182 (2005): 253–276.

Gregory, Paul R. *The Political Economy of Stalinism: Evidence from the Soviet Secret Archives*. New York: Cambridge University Press, 2003.

Gregory, Paul R. *Restructuring the Soviet Economic Bureaucracy*. New York: Cambridge University Press, 1990.

Grossman, Gregory. "Notes for a Theory of the Command Economy." *Soviet Studies* 15 (2) (1963): 101–123.

Grossman, Gregory. "The 'Second Economy' of the USSR." *Problems of Communism* 26 (5) (1977): 25–40.

Grossman, Gregory. "The Shadow Economy in the Socialist Sector of the USSR." In *The CMEA Five Year Plans (1981–1985) in a New Perspective: Planned and Non-Planned Economies*. Edited by Economics and Information Directorate. Brussels: North Atlantic Treaty Organization, 1982.

Grossman, Gregory, ed. *Studies in the Second Economy of Communist Countries: A Bibliography*. Berkeley: University of California Press, 1988.

Günther, Gotthard. "Cybernetics and Dialectical Materialism of Marx and Lenin." In *Computing in Russia: The History of Computer Devices and Information Technology Revealed*. Edited by Georg Trogemann, Alexander Nitussov, and Wolfgang Ernst, 317–332. Braunschweig: Vieweg, 2001.

Habermas, Jürgen. *The Structural Transformations of the Public Sphere: An Inquiry into a Category of Bourgeois Society*. Cambridge: MIT Press, 1989.

Hafner, Katie, and Matthew Lyon. *Where the Wizards Stay Up Late: The Origins of the Internet*. New York: Simon & Schuster, 1996.

Halpern, Orit. "Dreams for Our Perceptual Present: Archives, Interfaces, and Networks in Cybernetics." *Configurations* 13 (2007): 283–320.

Haraway, Donna. "A Cyborg Manifesto: Science, Technology, and Socialist-Feminism in the Late Twentieth Century." In *Simians, Cyborgs and Women: The Reinvention of Nature*. Edited by Donna Haraway. New York: Routledge, 1991.

Haraway, Donna, ed. *Simians, Cyborgs and Women: The Reinvention of Nature*. New York: Routledge, 1991.

Harrison, Mark. "Soviet Economic Growth since 1928: The Alternative Statistics of G. I. Khanin." *Europe-Asis Studies* 45 (1) (1993): 141–167.

Hayek, Friedrich. *The Road to Serfdom*. Chicago: University of Chicago Press, 1944.

Hayles, N. Katherine. *How We Became Posthuman: Virtual Bodies in Cybernetics, Literature, and Informatics*. Chicago: University of Chicago Press, 1999.

Hazard, John N. *Communists and Their Law: A Search for the Common Core of the Legal Systems of the Marxian Socialist States*. Chicago: University of Chicago, 1969.

Heidegger, Martin. *The Question concerning Technology and Other Essays*. Translated by William Lovitt. New York: Harper & Row, [1954] 1977.

Heims, Steve J. *The Cybernetics Group*. Cambridge: MIT Press, 1991.

Heims, Steve J. *John von Neumann and Norbert Wiener: From Mathematics to the Technologies of Life and Death*. Cambridge: MIT Press, 1982.

Hobsbawm, Eric. *The Age of Extremes: A History of the World, 1914–1991*. New York: Pantheon Books, 1994.

Hobsbawm, Eric. *How to Change the World: Reflections on Marx and Marxism*. New Haven: Yale University Press, 2011.

Hoffmann, David E. *The Dead Hand: The Untold Story of the Cold War Arms Race and Its Dangerous Legacy*. New York: Random House, 2009.

Hoffmann, Eric P., and Robbin F. Laird. *Technocratic Socialism: The Soviet Union in the Advanced Industrial Era*. Durham: Duke University Press, 1985.

Holloway, David. "Innovation in Science: The Case of Cybernetics in the Soviet Union." *Science Studies* 4 (1974): 299–337.

Holloway, David. "Physics, the State, and Civil Society in the Soviet Union." *Historical Studies in Physical and Biological Sciences* 100 (1) (1999): 173–193.

Honig, Bonnie. *Democracy and the Foreigner*. Princeton: Princeton University Press, 2003.

Honig, Bonnie, ed. *Feminist Interpretations of Hannah Arendt*. University Park: Pennsylvania State University Press, 1995.

Howe, R. H. "Max Weber's Elective Affinities." *American Journal of Sociology* 84 (1978): 366–385.

Hughes, Thomas. *Rescuing Prometheus: Four Monumental Projects That Changed the Modern World*. New York: Vintage, 2000.

Husbands, Phil, and Owen Holland. "The Ratio Club: A Hub of British Cyberneticists." In *The Mechanical Mind in History*. Edited by P. Husbands, O. Holland, and M. Wheeler, 91–148. Cambridge: MIT Press, 2008.

Isaacson, Walter. *The Innovators: How a Group of Inventors, Hackers, Geniuses, and Geeks Created the Digital Revolution.* New York: Simon & Schuster, 2014.

Jackson, Steven J. "Rethinking Repair." In *Media Technologies: Essays on Communication, Materiality and Society.* Edited by Tarleton Gillespie, Pablo Boczkowski, and Kirsten Foot, 221–240. Cambridge: MIT Press, 2013.

John, Nicholas, and Benjamin Peters. "Is the End Always Near? An Analysis and Comment on the End of Privacy, 1990–2012." Unpublished manuscript.

John, Richard. *Network Nation: Inventing American Telecommunications.* Cambridge: Harvard University Press, 2010.

Johnson, Daniel. *White King and Red Queen: How the Cold War Was Fought on the Chessboard.* Boston: Houghton Mifflin, 2008.

Johnson, Simon, Daniel Kaufmann, and Andrei Shleifer. "The Unofficial Economy in Transition." *Brookings Papers on Economic Activity* 2 (1997): 159–221.

Johnston, John. *The Allure of Machinic Life: Cybernetics, Artificial Life, and the New AI.* Cambridge, MA: MIT Press, 2008.

Joravsky, David. *Soviet Marxism and Natural Science, 1917–1932.* New York: Columbia University Press, 1971.

Joravsky, David. *The Lysenko Affair.* Chicago: University of Chicago Press, 1970.

Josephson, Paul R. *New Atlantis Revisited: Akademgorodok, the Siberian City of Science.* Princeton: Princeton University Press, 1977.

Josephson, Paul R. *Red Atom: Russia's Nuclear Power Program from Stalin to Today.* Pittsburg: University of Pittsburg Press, 2005.

Josephson, Paul R. *Totalitarian Science and Technology.* Atlantic Highlands, NJ: Humanities Press, 1996.

Jurcau, Nicolae. "Two Specialists in Cybernetics: Stefan Odobleja and Norbert Wiener, Common and Different Features." *Twentieth World Congress of Philosophy* (1998), accessed October 11, 2011, http://www.bu.edu/wcp/Papers/Comp/CompJurc.htm.

Kahnemann, Daniel. *Thinking Fast and Slow.* New York: Farrar, Straus, and Giroux, 2011.

Kahneman, Daniel and Amos Tversky. "Prospect Theory: An Analysis of Decisions under Risk," *Econometrica* 47 (2) (1979): 263–291.

Kahneman, Daniel and Amos Tversky, eds., *Choices, Values and Frames.* New York: Cambridge University Press and Russell Sage Foundation, 2000.

Kapitonova, Yulia O., and Aleksandr A. Letichevsky, *Paradigmi i idei akademika V. M. Glushkova* [*Paradigms and ideas of academician V. M. Glushkov*]. Kiev: Naukova Dumka, 2003.

Kapp, William. *The Foundations of Institutional Economics.* New York: Routledge, 2011.

Kassel, Simon. *Soviet Cybernetics Research: A Preliminary Study of Organizations and Personalities.* Santa Monica: RAND, 1971.

Kay, Lily E. "Cybernetics, Information, Life: The Emergence of Scriptural Representations of Heredity." *Configurations* 5 (1) (1997): 23–91.

Keane, John. *The Life and Death of Democracy.* New York: Simon & Schuster, 2009.

Kelly, Kevin. *Out of Control: The New Biology of Machines, Social Systems, and the Economic World.* 4th ed. New York: Addison Wesley, 2004.

Kharkevich, Aleksandr. "Informatsia i tekhnika" ["Information and technology"]. *Kommunist* 17 (1962): 93–102.

Khinchin, Aleksandr Ya. "Teoria prosteishego potoka" ["Simple stream theory."] *Trudy Matematicheskogo Instituta Steklov* 49 (1955): 3–122.

Khrushchev, Nikita. "Concluding Speech." *Doklad na XXII S'esde KPSS* (October 18, 1961). Accessed March 19, 2010, http://www.archive.org/details/DocumentsOfThe2 2ndCongressOfTheCpsuVolI.

Khrushchev, Sergei N. *Nikita Khrushchev and the Creation of a Superpower.* University Park: Pennsylvania State University Press, 2000.

Kim, Byung-Yeon. "Informal Economy Activities of Soviet Households: Size and Dynamics." *Journal of Comparative Economics* 31 (3) (2003): 532–551.

Kitov, Anatoly. "Chelovek, kotoryi vynes kibernetiku iz sekretnoi biblioteki" ["The Man Who Brought Cybernetics out of a Secret Library"]. *Komp'iuterra* 18 (43) (1996): 44–45.

Kitov, Anatoly. "Perecen' osnovnykh nauchnykh trudov" ["List of basic scientific works]. Accessed April 15, 2015, http://www.kitov-anatoly.ru/naucnye-trudy/ perecen-osnovnyh-naucnyh-trudov.

Kitov, Anatoly. *Electronnie tsifrovie mashini* [Electronic cipher machines]. Moscow: Radioeletronika Nauka, 1956.

Kitov, Anatoly. "Kibernetika i upravlenie narodnym khoziastvom" ["Cybernetics and the management of the national economy"], *Kibernetiku—na sluzhbu Kommunism.* Edited by Aksel' Berg, 207–216. Moscow: Gosekkergoizdat, 1961.

Kitov, Anatoly. "Rol' akademika A. I. Berga v razvitii vyichislitel'noi tekhniki I avtomatizirovannikh system upravleniya" ["The role of the academician A. I. Berg in developing computing technology and automated systems of management"]. In *Put' v bol'shuyu nauku: Akademik Aksel' Berg.* Moscow: Nauka, 1988. Accessed March 20, 2010, http://www.computer-museum.ru/galglory/berg3.htm.

Kitov, Viktor A., E. N. Filinov, and L. G. Chernyak. "Anatoly Ivanovich Kitov." Accessed May 19, 2010, http://www.computer-museum.ru/galglory/kitov.htm.

Kitov, Vladimir A., and Valery V. Shilov. "Anatoly Kitov: Pioneer of Russian Informatics." In *History of Computing: Learning from the Past*, vol. 325. Edited by Arthur Tatnall, 80–88. New York: Springer, 2010.

Kline, Ronald R. *The Cybernetics Moment, Or Why We Call Our Age the Information Age.* Baltimore, MD: Johns Hopkins University Press, 2015.

Kline, Ronald R. "What Is Information Theory a Theory Of? Boundary Work among Scientists in the United States and Britain during the Cold War." In *The History and Heritage of Scientific and Technical Information Systems: Proceedings of the 2002 Conference, Chemical Heritage Foundation.* Edited by W. Boyd Rayward and Mary Ellen Bowden. Medford, NJ: Information Today, 2004.

Kline, Ronald R. "Where Are the Cyborgs in Cybernetics?" *Social Studies of Science* 39 (3) (2009): 331–362.

Kohler, Robert E., and Kathryn M. Olesko. "Introduction: Clio Meets Science: The Challenges of History." *Osiris* 27 (1) (2012): 4–6.

Kornai, János. *The Socialist System: The Political Economy of Communism.* Princeton: Princeton University Press, 1992.

Kozulin, Alex. *Vygotsky's Psychology: A Biography of Ideas.* Cambridge: Harvard University Press, 1990.

Kranzberg, Melvin. "Technology and History: 'Kranzberg's Laws.'" *Technology and Culture* 27 (3) (1986): 544–560.

Kreiss, Daniel, Megan Finn, and Fred Turner. "The Limits of Peer Production: Some Reminders from Max Weber for the Network Society." *New Media and Society* 13 (2) (2011): 243–259.

Krementsov, Nikolai. *A Martian Stranded on Earth: Alexander Bogdanov, Blood Transfusions, and Proletarian Science.* Chicago: University of Chicago Press, 2011.

Krementsov, Nikolai. *Stalinist Science.* Princeton: Princeton University Press, 1996.

Kuteinikov, Aleksei. "Akademik V. M. Glushkov i proekt sozdaniya printsipial'no novoi (avtomatizirovannoi) sistemi upravleniya sovetskoi ekonomikoi v 1963–1965 gg." ["Academician V. M. Glushkov and creating the principally new project of an (automated) system of management for the Soviet economy in 1963-1965"]. *Ekonomicheskaya istoriya*, 139–156. Moscow: Trudi istoricheskogo faku'teta MGU 15, 2011.

Kuteinikov, Aleksei. "Pervie proekti avtomatizatsii upravleniya sovetskoi planovoi ekonomiki v kontse 1950-x i nachale 1960-x gg.—'elektronnyi sotsializm'?" ["First projects in the automation of management of Soviet planned economics from the endof the 1950s to the beginning of the 1960s—"electronic socialism?"]. *Ekonomicheskaya istoriya*, 124–138. Moscow: Trudi istoricheskogo faku'teta MGU 15, 2011.

Kuteinikov, Aleksei Viktorovich. "Proekt Obshchegosudarstvennoi avtomatizirovan-noi sistemi upravleniya sovetskoi ekonomikoi (OGAS) i problem ego realizatsii v 1960–1980-x gg." ["The all-state automated system of management of Soviet economics project (OGAS) and the problem of its realization in 1960-1980"]. Ph.D. diss., Moscow State University, 2011.

Lafontaine, Céline. "The Cybernetic Matrix of 'French Theory.'" *Theory, Culture and Society* 24 (2007): 27–46.

LaFontaine, Céline. *L'empire cybernétique: des machines à penser à la pensée machine.* Paris: Seuil, 2004.

Landis, E. M., and I. M. Yaglom. "About Aleksandr Semenovich Kronrod." *Uspekhi Matematicheskikh Nauk* 56 (5), 341 (2001): 191–201. Accessed April 15, 2015, http://www.mathnet.ru/links/1e483992e9f2c42fda4390d0116737a3/rm448.pdf.

Latour, Bruno. *Reassembling the Social: An Introduction to Actor-Network-Theory.* Oxford: Oxford University Press, 2005.

Latour, Bruno. *Science in Action: How to Follow Scientists and Engineers through Society.* Milton Keynes: Open University Press, 1987.

Latour, Bruno. "Technology Is Society Made Durable." In *A Sociology of Monsters: Essays on Power, Technology and Domination.* Sociological Review Monograph no. 38. London: Routledge, 1991.

Law, John, ed. *A Sociology of Monsters: Essays on Power, Technology and Domination.* Sociological Review Monograph no. 38. London: Routledge, 1991.

Lax, David Alexander. *Libermanism and the Kosygin Reform.* Charlottesville: University of Virginia Press, 1991.

Lebedev, Sergei, Lev Dashevsky, and Ekaterina Shkabara. "Malaya elektronnaya shchyotnyaya mashina," ["The small electronic calculating machine"] 1952. Accessed April 15, 2015, http://it-history.ru/images/a/af/SALebedev_MESM.pdf.

Ledeneva, Alena V. *Can Russia Modernize? Sistema, Power Networks and Informal Governance.* New York: Cambridge University Press, 2013.

Ledeneva, Alena V. *Russia's Economy of Favors: Blat, Networking and Information Exchange.* New York: Cambridge University Press, 1998.

Leontief, Wassily W. *Studies in the Structure of the American Economy.* New York: Oxford University Press, 1953.

Leslie, Stuart W. *The Cold War and American Science: The Military-Industrial-Academic Complex.* New York: Columbia University Press, 1993.

Lessig, Lawrence. *Code and Other Laws of Cyberspace.* New York: Basic Books, 1999.

Lettvin, Jerome Y., Humberto R. Maturana, Warren S. McCulloch, and Walter Pitts. "What the Frog's Eye Tells the Frog's Brain." *Proceedings of the IRE,* 47 (11), (November

1959). Reprinted in *The Collected Works of Warren S. McCulloch*, vol. 4. Edited by Rook McCulloch, 1161–1172. Salinas, CA: Intersystems Publications, 1989.

Liberman, Evsei G. "Plans, Profits and Bonuses." In *The Liberman Discussion: A New Phase in Soviet Economic Thought*. Edited by M. E. Sharpe. White Plains, NY: International Arts and Science Press, 1965. Originally published in *Pravda*, September 9, 1962.

Licklider, J.C.R. "Man-Machine Symbiosis." *IRE Transactions on Human Factors in Electronics*, HFE-1 (1960): 4–11. Accessed July 25, 2013, http://groups.csail.mit.edu/medg/people/psz/Licklider.html.

Liebniz, Gottfried Wilhelm Freiherr. "The Art of Discovery" (1685). In *Leibniz: Selections*. Edited by Philip P. Wiener, 50–58. New York: Charles Scribner's Sons, 1951.

Light, Jennifer. *From Warfare to Welfare: Defense Intellectuals and Urban Problems in Cold War America*. Baltimore: John Hopkins University Press, 2003.

Light, Jennifer S. "When Computers Were Women." *Technology and Culture* 40 (3) (1999): 455–483.

Lipset, David. *Gregory Bateson: The Legacy of a Scientist*. New York: Prentice Hall, 1980.

Liu, Lydia. "The Cybernetic Unconscious: Rethinking Lacan, Poe, and French Theory." *Critical Inquiry* 36 (2010): 288–320.

Lovell, Steven. *The Soviet Union: A Very Short Introduction*. New York: Oxford University Press, 2009.

Machlup, Fritz. *The Production and Distribution of Knowledge in the United States*. Princeton: Princeton University Press, 1962.

Mackrasis, Kristie. *Seduced by Secrets: Inside the Stasi's Spy-Tech World*. New York: Cambridge University Press, 2014.

Malcolm, D. G. "Review of Cybernetics at [sic] Service of Communism, vol. 1." *Operations Research* 11 (1963): 1012.

Malinovsky, Boris Nikolaevich. *Istoriaa vychislitel'noi tekhniki*. Kiev: Kit, 1995. Accessed April 15, 2015, http://lib.ru/MEMUARY/MALINOWSKIJ/0.htm.

Malinovsky, Boris Nikolaevich. *Pioneers of Soviet Computing*. Edited by Anne Fitzpatrick. Translated by Emmanuel Aronie. Accessed April 15, 2015, http://www.sigcis.org/files/SIGCISMC2010_001.pdf.

Malinovsky, Boris Nikolaevich. *Vechno Khranit [Store Eternally]*. Kiev: Gorobets, 2007.

Marolla, Paul A., John V. Arthur, Rodrigo Alvarez-Icaza, Andrew S. Cassidy, Jun Sawwada, Filipp Akopyan, Bryan L. Jackson, Nabil Imam, Chen Guo, Yutaka Nakamura, Bernard Brezzo, Ivan Vo, Steven K. Esser, Rathinakumar Appuswamy, Brian Taba, Arnn Amir, Myron D. Flickner, William P. Risk, Rajit Manohar, and Dharmendra S. Modha. "A Million Spiking-Neuron Integrated Circuit with a Scalable Communication

Network and Interface." *Science* 8 (345), 6197 (2014): 668–673. Accessed April 15, 2015, http://www.sciencemag.org/content/345/6197/668.

Mars, Gerald, and Altman, Yochanan. *Private Enterprise in the USSR: The Case of Soviet Georgia*. Aldershot: Gower Press, 1987.

Marx, Karl, and Friedrich Engels. *Karl Marx and Frederick Engels: Selected Correspondence*. Moscow: Progress, 1975.

Masani, Pesi R. *Nobert Wiener, 1894–1964*. Boston: Birkhäuser Verlag, 1990.

Maturana, Humberto, and Francisco Valera, *Autopoiesis and Cognition: The Realization of the Living*. Boston: Reidel, 1980.

Mayr, Otto. *Authority, Liberty, and Automatic Machinery in Early Modern Europe*. Baltimore: Johns Hopkins University Press, 1989.

McCarthy, John. "Chess as the Drosophila of AI." Accessed April 15, 2015, http://jmc.stanford.edu/articles/drosophila/drosophila.pdf.

McCulloch, Warren S. "A Heterarchy of Values Determined by the Topology of Nervous Nets." *Bulletin of Mathematical Biophysics* 7 (1945): 89–93.

McCulloch, Warren S., and Walter Pitts. "A Logical Calculus of Ideas Immanent in Nervous Activity," *Bulletin of Mathematical Biophysics* 5 (1943):115–133.

McDonald, Christopher Felix. "Building the Information Society: A History of Computing as a Mass Medium." Ph.D. diss., Princeton University, 2011.

McLuhan, Marshall. *Understanding Media: The Extensions of Man*. New York: McGraw-Hill, 1964.

McMahon, Robert. *The Cold War: A Very Short Introduction*. New York: Oxford University Press, 2003.

McNeely, Ian F., with Lisa Wolverton. *Reinventing Knowledge: From Alexandria to the Internet*. New York: Norton, 2008.

Medema, Steven G. *The Hesitant Hand: Taming Self-Interest in the History of Economic Ideas*. Princeton: Princeton University Press, 2010.

Medina, Eden. *Cybernetic Revolutionaries: Technology and Politics in Allende's Chile*. Cambridge: MIT Press, 2011.

Medina, Eden, ed. *Beyond Imported Magic: Essays on Science, Technology, and Society in Latin America*. Cambridge: MIT Press, 2014.

Medvedev, Zhores. *Soviet Science*. New York: Norton, 1978.

Meerovich, Eduard A. "Obsuzhdenie doklada professor A. A. Liapunova 'Ob ispol'zovanii matematicheskikh mashin v logicheskikh tseliakh" ["Discussion of Professor A. A. Liapunov's report, 'On the use of mathematical machines in logical goals.'"] (1954), in *Ocherki istorii informatiki*. Edited by D. Pospelov and Ya. Fet.

Novosibirsk: Nauchnyi Tsentr Publikatsii RAS, 1998. Accessed March 20, 2010, http://ssd.sscc.ru/PaCT/history/early.html.

Michailova, Snejina, and Verner Worm. "Personal Networking in Russia and China: Blat and Guanxi." *European Management Journal* 21 (4) (2003): 509–519.

Mindell, David. *Between Human and Machine: Feedback, Control, and Computing before Cybernetics.* Baltimore: Johns Hopkins University Press, 2002.

Mindell, David, Jerome Segal, and Slava Gerovitch. "From Communications Engineering to Communications Science: Cybernetics and Information Theory in the United States, France, and the Soviet Union." In *Science and Ideology: A Comparative History.* Edited by Mark Walker, 66–96. New York: Routledge, 2003.

Mirowski, Philip. *Machine Dreams: Economics Becomes a Cyborg Science.* New York: Cambridge University Press, 2001.

Moore, Gordon E. "Cramming More Components onto Integrated Circuits." *Electronics* 38 (8) (1965): 114–117.

Mosco, Vincent. *To the Cloud: Big Data in a Turbulent World.* New York: Paradigm Publishers, 2014.

Mumford, Lewis. *Technics and Civilization.* New York: Harvest Book, 1934.

Najafi, Sina, and Peter Galison. "The Ontology of the Enemy: An Interview with Peter Galison," *Cabinet* 12 (2003), accessed April 10, 2015: http://cabinetmagazine.org/issues/12/najafi2.php.

Nemchinov, Vasily S. *O dalneishem sovershenstvovanii planirovaniya i upravleniya narodnym khozyaistvom [On the further improvement of the planning and management of the national economy].* Moscow: Ekonomika, 1964.

Nemchinov, Vasily S. "Sotsialisticheskoe khozyaistvovanie i planirovanie proizvodstva" ["Socialist economic management and planning of production"]. *Kommunist* 5 (1964).

Nielsen, Rasmus Kleis. "Democracy." In *Digital Keywords: A Vocabulary for Information Society and Culture.* Edited by Benjamin Peters, forthcoming. Princeton: Princeton University Press, 2016. Accessed April 15, 2015. http://culturedigitally.org/2014/05/democracy-draft-digitalkeywords/.

North, Douglass C. *Institutions, Institutional Change and Economic Performance.* New York: Cambridge University Press, 1998.

Nove, Alec. *An Economic History of the USSR, 1917–1991.* 3rd ed. New York: Penguin Group, 1992.

Odom, William E. *The Collapse of the Soviet Military.* New Haven: Yale University Press, 1998.

O'Hearn, Dennis. "The Consumer Second Economy: Size and Effects." *Soviet Studies* 32 (2) (April 1980): 218–234.

O'Neill, Judy. "Interview with Paul Baran." Charles Babbage Institute, OH 182, Menlo Park, CA, March 5, 1990. Accessed April 15, 2015, http://www.gtnoise.net/classes/cs7001/fall_2008/readings/baran-int.pdf.

Ong, Walter. *Orality and Literacy*. New York: Routledge, 1972.

O'Shea, Michael. *The Brain: A Very Short Introduction*. New York: Oxford University Press, 2005.

Osokina, Elena. *Our Daily Bread: Socialist Distribution and the Art of Survival in Stalin's Russia, 1927–1941*. New York: Routledge, 2003.

Pallo, Gabor. "The Hungarian Phenomenon in Israeli Science." *Bulletin of the History of Chemistry* 25 (1) (2000): 35–42.

Paloczi-Horvath, George. *Khrushchev: The Making of a Dictator*. New York: Little, Brown, 1960.

Pavlov, Ivan. *Conditioned Reflexes: An Investigation of the Physiological Activity of the Cerebral Cortex*. Translated and edited by G. V. Anrep. London: Oxford University Press, 1927.

Peirce, Charles Sanders. "How to Make Our Ideas Clear." In *The Essential Peirce*, Vol. 1, 334–351. Bloomington: Indiana University Press, 1992–1999.

Peters, Benjamin. "And Lead Us Not into Thinking the New Is New: A Bibliographic Case for New Media History." *New Media and Society* 11 (1–2) (2009): 13–30.

Peters, Benjamin. "Toward a Genealogy of a Cold War Communication Science: The Strange Loops of Leo and Norbert Wiener." *Russian Journal of Communication* 5 (1) (2013): 31–43.

Peters, Benjamin, and Deborah Lubken. "New Media in Crises: Discursive Instability and Emergency Communication." In *The Long History of New Media*. Edited by David W. Park, Nicholas W. Jankowski, and Steve Jones, 193–209. New York: Peter Lang, 2011.

Pias, Claus. "Analog, Digital, and the Cybernetic Illusion." *Kybernetes* 34 (3–4) (2005): 543–550.

Pias, Claus, ed. *Cybernetics—Kybernetik 2: The Macy-Conferences 1946–1953*. Berlin: Diaphanes, 2004.

Pickering, Andrew. *The Cybernetic Brain: Sketches of Another Future*. Chicago: University of Chicago Press, 2010.

Pierce, J. R. "The Early Days of Information Theory," *IEEE Transactions on Information Theory* 19 (1) (1973): 3–8.

Pipes, Richard. *Russian Conservatism and its Critics: A Study in Political Culture.* New Haven: Yale University Press, 2005.

Poletaev, Igor' A. "K opredeleniiu poniatiia 'informatsiia'" ["Toward a definition of 'information'"] In *Issledovanniia po kibernetike.* Edited by A. Lyapunov, 210–214. Moscow: Sovetskoe radio, 1970.

Poletaev, Igor' A. "O matematicheskom modelirovnanii." *Problemy kibernetiki* 27 (1973): 145–149.

Pollock, Ethan. *Stalin and the Soviet Science Wars.* Princeton: Princeton University Press, 2006.

Pooley, Jefferson. "An Accident of Memory: Edward Shils, Paul Lazarsfeld and the History of American Mass Communication Research," Ph.D. diss., Columbia University, New York, 2006.

Portes, Alejandro, Manuel Castells, and Lauren A. Benton, eds. *The Informal Economy: Studies in Advanced and Less Developed Countries.* Baltimore: Johns Hopkins University Press, 1989.

Pospelov, D., and Ya. Fet, eds. *Ocherki istorii informatiki v Russii [Essays in the History of Informatics in Russia].* Novosibirsk: Nauchnyi Tsentr Publikatsii RAS, 1998. Accessed March 20, 2010, http://ssd.sscc.ru/PaCT/history/early.html.

Priest, George L. "The Ambiguous Moral Foundations of the Underground Economy." Faculty Scholarship Series, Paper 626 (1995). Accessed April 15, 2015, http://digitalcommons.law.yale.edu/cgi/viewcontent.cgi?article=1625&context=fss_papers.

Raeff, Marc. *The Well-Ordered Police State: Social and Institutional Change through Law in the Germanies and Russia, 1600–1800.* New Haven: Yale University Press, 1983.

Raymond, Eric. *The Cathedral and the Bazaar: Musings on Linux and Open Source by an Accidental Revolutionary.* New York: O'Reilly, 1999.

Reese, Roger R. *The Soviet Military Experience: A History of the Soviet Army, 1917–1991.* New York: Routledge, 2000.

Remnick, David. "Soviet Union's Shadow Economy: Bribery, Barter, and Black Market." *Seattle Times,* September 22, 1990. Accessed April 15, 2015, http://community.seattletimes.nwsource.com/archive/?date=19900922&slug=1094485.

Ries, Nancy. *Russian Talk: Culture and Conversation during Perestroika.* Ithaca: Cornell University Press, 1997.

Roll-Hansen, Nils. *The Lysenko Effect: The Politics of Science.* Amherst, NY: Humanity Books, 2005.

Rosenblueth, Arturo, Norbert Wiener, and Julian Bigelow. "Behavior, Purpose, and Teleology." *Philosophy of Science* 10 (1943): 18–24.

Rudnev, K. N. "Vyichislitel'naya tekhnika v narodnom khozyyaistve" ["Computing technology in the national economy"]. *Izvestiya*, September 4, 1963.

Ryavec, Karl W. *Russian Bureaucracy: Power and Pathology*. New York: Rowman & Littlefield, 2005.

Salus, Peter H., ed. *The ARPANET Sourcebook*. Charlottesville, VA: Peer-to-Peer Communications LLC, 2008.

Samuels, Richard J. *"Rich Nation, Strong Army": National Security and the Technological Transformation of Japan*. Ithaca: Cornell University Press, 1994.

Schatzberg, Eric. "Technik Comes to America: Changing Meanings of Technology before 1930." *Technology and Culture* 47 (2006): 486–511.

Schelling, Thomas C. *Micromotives and Macrobehavior*. New York: Norton, 1978.

Schroeder, Gertrude E. "Organizations and Hierarchies: The Perennial Search for Solutions." In *Reorganization and Reform in the Soviet Economy*. Edited by Susan J. Linz and William Moskoff, 3–22. New York: Sharpe, 1988.

Schroeder, Gertrude E. "The Soviet Economy on a Treadmill of Reforms." *Soviet Economy in a Time of Change*. Washington, DC: U.S. Congress Joint Economic Committee, 1979.

Segal, Howard P. *Technology and Utopia*. Washington, DC: American Historical Association, 2006.

Segal, Jérôme. *Le zéro et le un: histoire de la notion scientifique d'information*. Paris: Syllepse, 2003.

Segal, Jérôme. "L'introduction de la cybernétique en R.D.A. rencontres avec l'idéologie marxiste," *Science, Technology and Political Change: Proceedings of the Twentieth International Congress of History of Science* (Liège, July 20–26, 1997) (Brepols: Turnhout, 1999), 1: 67–80.

Sellen, Abigail, and Richard Harper. *The Myth of the Paperless Office*. Cambridge: MIT Press, 2003.

Shabad, Theodore. "Khrushchev Says Missile Can 'Hit a Fly' in Space." *New York Times*, July 17, 1962.

Shannon, Claude E. "A Mathematical Theory of Communication." *Bell Systems Technical Journal* 27 (1948): 379–423, 623–656.

Shannon, Claude E. "The Bandwagon," *IRE Transactions on Information Theory* 2 (1) (1956): 3.

Shipler, David. *Russia: Broken Idols, Solemn Dreams*. New York: Times Books, 1983.

Shirky, Clay. *Cognitive Surplus: Creativity and Generosity in a Connected Age*. New York: Penguin Books, 2010.Shklovsky, Viktor. "Art as Technique" (1917). In *Russian*

Formalist Criticism: Four Essays. Edited by Lee T. Lemon and Marion J. Reiss, 3–24. Lincoln: University of Nebraska Press, 1965.

Simis, Konstantin. *USSR: The Corrupt Society—The Secret World of Soviet Capitalism.* New York: Simon & Schuster, 1982.

Simmel, Georg. *Philosophie des Geldes.* Leipzig: Duncker & Humblot, 1900.

Smolinski, Leon. "What Next in Soviet Planning?" *Foreign Affairs* 42 (3) (1964): 603–613.

Snyder, Timothy. *Bloodlands: Europe between Hitler and Stalin.* New York: Basic Books, 2012.

Sobolev, Sergei L, Anatolii I. Kitov, and Aleksei A. Lyapunov. "Osnovnye cherty kibernetiki" ["Basic Features of Cybernetics"]. *Voprosy filosofii* 4 (1955): 141.

Soll, Jacob. *The Information Master: Jean-Baptiste Colbert's Secret State Intelligence System.* Ann Arbor: University of Michigan Press, 2009.

Solove, Daniel J. "A Taxonomy of Privacy." *University of Pennsylvania Law Review* 154 (3) (2006): 477–560.

Soyfer, Valery N. *Lysenko and the Tragedy of Soviet Science.* New Brunswick, NJ: Rutgers University Press, 1994.

Spencer, Herbert. *The Principles of Sociology.* 3rd ed. New York: Westminster, [1876] 1896.

Spufford, Francis. *Red Plenty.* Minneapolis: Greywolf Press, 2010.

Stalin, Iosif V. *Voprosy leninizma* [*Questions of Leninism*]. 11th ed. Moscow: Gospolitizdat, 1951.

Stark, David. *The Sense of Dissonance: Accounts of Worth in Economic Life.* Princeton: Princeton University Press, 2009.

Starr, S. Frederick. *Red and Hot: The Fate of Jazz in the Soviet Union, 1917–1991.* New York: Oxford University Press, 1994.Stillman, Boris. *Linguistic Geometry.* New York: Springer, 2000.

Streeter, Thomas. *The Net Effect: Romanticism, Capitalism, and the Internet.* New York: New York University Press, 2011.

Taubman, William, Sergei Khrushchev, and Abbott Gleason. *Nikita Khrushchev.* New Haven: Yale University Press, 2000.

Thiel, Peter. *Zero to One: Notes on Startups, or How to Build the Future.* New York: Crown Business, 2014.

Timofeev, Lev. *Soviet Peasants: Or the Peasants' Art of Starving.* New York: Telos Books, 1985.

Tofts, Darren, Annemarie Jonson, and Alessio Cavallaro, eds. *Prefiguring Cyberculture: An Intellectual History*. Cambridge: MIT Press, 2002.

Trotsky, Leon. *Platform of the Joint Opposition* (1927). London: New Park Publications, 1973.

Tuchkov, Vladmir Yakovlevich. *Pervoprokhodets tsivorogo materika: Anatoliyi Ivanovich Kitov* [*Earliest explorer of the digital continent: Anatoly Ivanovich Kitov*] Moscow: FGBO VPO, 2014.

Tucker, Robert C., ed. *The Marx-Engels Reader, 2nd Edition*. New York: W. W. Norton, 1978.

Turner, Fred. *From Counterculture to Cyberculture*. Chicago: University of Chicago Press, 2006.

Umpleby, Stuart. "A History of the Cybernetics Movement in the United States." *Journal of the Washington Academy of Sciences* 91 (2005): 54–66.

Valera, Francisco. *The Tree of Knowledge: The Biological Roots of Human Understanding*. Boston: Shambhala Press, 1987.

Valera, Francisco with Evan Thompson and Eleanor Rosch. *The Embodied Mind: Cognitive Science and Human Experience*. Cambridge: MIT Press, 1991.

Veblen, Thorsten. *The Engineers and the Price System*. New York: Huebsch, 1921.

Veblen, Thorsten. "Why Is Economics Not an Evolutionary Science?" *Quarterly Journal of Economics* 12 (1898): 373–397.

Volkov, Solomon. *The Magical Chorus: A History of Russian Culture from Tolstoy to Solzhenitsyn*. New York: Knopf, 2008.

Volkov, Vadim. *Violent Entrepreneurs: The Use of Force in the Making of Russian Capitalism*. Ithica: Cornell University Press, 2002.

Von Eschen, Penny. *Satchmo Blows Up the World: Jazz Ambassadors Play the Cold War*. Cambridge: Harvard University Press, 2004.

Von Hilger, Philip. *War Games: A History of War on Paper*. Cambridge: MIT Press, 2012.

Von Neumann, John. "Can We Survive Technology?" *Fortune* (June 1955): 106–108, 151–152.

Von Neumann, John. *The Computer and the Brain*. 2nd ed. New Haven: Yale University Press, [1958] 2000.

Voytek, Bradley. "Are There Really as Many Neurons in the Human Brain as Stars in the Milky Way?" *Nature*, May 20, 2013. Accessed April 15, 2015, http://www.nature.com/scitable/blog/brain-metrics/are_there_really_as_many.

Vucinich, Alexander. *Empire of Knowledge: The Academy of Sciences of the USSR (1917–1970)*. Berkeley: University of California Press, 1984.

Wädekin, Karl Eugen. *The Private Sector in Soviet Agriculture*. Berkeley: University of California Press, 1973.

Wark, McKenzie. *Molecular Red: A Theory for the Anthropocene*. New York: Verso, 2015.

Weber, Max. *The Protestant Ethic and the Spirit of Capitalism*. Translated by Talcott Parsons and Anthony Giddens. Boston: Unwin Hyman, 1930 [originally published in 1905].

Weintraub, Jeff. "The Theory and Politics of the Public/Private Distinction." *Public and Private in Thought and Practice: Perspectives on a Grand Dichotomy*. Edited by Jeff Weintraub and Krishan Kumar, 8–16. Chicago: University of Chicago Press, 1997.

Wiener, Norbert. *Cybernetics, or Control and Communication in the Animal and the Machine*. Cambridge: MIT Press, 1948.

Wiener, Norbert. *God and Golem, Inc.: A Comment on Certain Points Where Cybernetics Impinges on Religion*. Cambridge: MIT Press, 1964.

Wiener, Norbert. *The Human Use of Human Beings; Cybernetics and Society*. Boston: Houghton Mifflin, 1950.

Wiener, Norbert. "Obschestvo i nauka." *Voprosi Filosofiii* 7 (1961), Reprinted in *[Massachusetts Institute of] Technology Review* 63 (July 1961), 49–52.

Williams, Raymond. "Base and Superstructure in Marxist Cultural Theory." *New Left Review* 1 (82) (1973): 1–13.

Williams, Raymond. "Culture Is Ordinary." In *Resources of Hope: Culture, Democracy, Socialism*, 3–14. London: Verso.

Williams, Raymond. *Television: Technology and Cultural Form*. London: Wesleyan University Press, 1974.

Wolfe, Audra J. "The Cold War Context of the Golden Jubilee, or, Why We Think of Mendel as the Father of Genetics." *Journal of the History of Biology* 45 (3) (2012): 389–414.

Wolfe, Audra J. *Competing with the Soviets: Science, Technology, and the State in Cold War America*. Baltimore: Johns Hopkins University Press, 2013.

Wu, Tim. *The Master Switch: The Rise and Fall of Information Empires*. New York: Atlantic Books, 2010.

Zittrain, Jonathan. *The Future of the Internet, and How to Stop It*. New Haven: Yale University Press, 2008.

Zubok, Vladislav. *Zhivago's Children: The Last Russian Intelligentsia*. Cambridge: Harvard University Press, 2009.

Index

Printed in the United States
by Baker & Taylor Publisher Services

Printed in the United States
by Baker & Taylor Publisher Services